T0184732

Cultural Spaces, Production and Consumption

This book explores the concept of cultural spaces, their production and how they are experienced by different users. It explores this concept and practice from formal and informal arts and heritage sites, festivals and cultural quarters – to the production of digital, fashion and street art, and social engagement through cultural mapping and site-based artist collaborations with local communities.

It offers a unique take on the relationship between cultural production and consumption through an eclectic range of cultural space types, featuring examples and case studies across cultural venues, events and festivals, and cultural heritage – and their usage. Cultural production is also considered in terms of the transformation of cultural and digital-creative quarters and their convergence as visitor destinations in city fringe areas, to fashion spaces, manifested through museumification and fashion districts. The approach taken is highly empirical supported by a wide range of visual illustrations and data, underpinned by key concepts, notably the social production of space, cultural rights and everyday culture, which are both tested and validated through the original research presented throughout.

This book will appeal to students and researchers in human geography, arts and museum management, cultural policy, cultural studies, architecture and town planning. It will also be useful for policymakers and practitioners from local and city government, government cultural agencies and departments, architects and town planners, cultural venues, arts centres, museums, heritage sites, and artistic directors/programmers.

Graeme Evans is Emeritus Professor of Creative and Cultural Economy at the University of the Arts London. He served as founding director and professor at the Cities Institute, London, and the Centre for Urban & Euregional Studies, Maastricht University, and held the posts of Professor of Design Cultures and the director of the Art & Design Research Institute, Middlesex University, and Professor of Design at Brunel University. Key publications include *Cultural Planning: An Urban Renaissance?* (Routledge), *Designing Sustainable Cities* and *Mega-Events: Placemaking, Regeneration and City Regional Development* (Routledge), and over a hundred chapters in edited books and journal articles. He has undertaken numerous grant-funded research projects for UK research councils, the European Commission and commissioned research studies on cultural policy for the Department for Culture Media and Sport (DCMS), Arts Council England, Historic England, HM Treasury, Council of Europe, UNESCO and the OECD; regional development agencies, including Creative London, Toronto Metro and Limburg Province, the Netherlands; and international cultural development agencies in South Korea, China, Canada and Sweden. Prior to academe, he was a musician, co-director of an arts centre in North London and director of the London Association of Arts Centres.

Cultural Spaces, Production and Consumption

Graeme Evans

Routledge
Taylor & Francis Group

LONDON AND NEW YORK

Designed cover image: © Graeme Evans

First published 2024
by Routledge
4 Park Square, Milton Park, Abingdon, Oxon OX14 4RN

and by Routledge
605 Third Avenue, New York, NY 10158

Routledge is an imprint of the Taylor & Francis Group, an informa business

© 2024 Graeme Evans

The right of Graeme Evans to be identified as author of this work has been
asserted in accordance with sections 77 and 78 of the Copyright, Designs
and Patents Act 1988.

All rights reserved. No part of this book may be reprinted or reproduced or
utilised in any form or by any electronic, mechanical, or other means, now
known or hereafter invented, including photocopying and recording, or in
any information storage or retrieval system, without permission in writing
from the publishers.

Trademark notice: Product or corporate names may be trademarks or
registered trademarks, and are used only for identification and explanation
without intent to infringe.

British Library Cataloguing-in-Publication Data
A catalogue record for this book is available from the British Library

Library of Congress Cataloging-in-Publication Data
Names: Evans, Graeme, author.
Title: Cultural spaces, production and consumption / Graeme Evans.
Description: First edition. | New York : Routledge, 2024. | Includes
    bibliographical references and index. | Contents: Introduction—A Place
    for the Arts—Events and Festivals—Cultural Heritage—Cultural and
    Creative Quarters—Digital Cultural Space—Fashion Spaces—Graffiti
    and Street Art—Socially Engaged Practice and Cultural Mapping.
Identifiers: LCCN 2023041906 (print) | LCCN 2023041907 (ebook) |
    ISBN 9781032106823 (hbk) | ISBN 9781032106830 (pbk) |
    ISBN 9781003216537 (ebk)
Subjects: LCSH: Heritage tourism. | Culture and tourism.
Classification: LCC G156.5.H47 E83 2024 (print) | LCC G156.5.H47
    (ebook) | DDC 338.4/791—dc23/eng/20231106
LC record available at https://lccn.loc.gov/2023041906
LC ebook record available at https://lccn.loc.gov/2023041907

ISBN: 978-1-032-10682-3 (hbk)
ISBN: 978-1-032-10683-0 (pbk)
ISBN: 978-1-003-21653-7 (ebk)

DOI: 10.4324/9781003216537

Typeset in Times New Roman
by Apex CoVantage, LLC

On a personal note, special thanks are made to Claudia for support and image editing and to Edie, Robbie and Isla for whom cultural spaces will continue to play an important part in their lives.

Contents

Figures

Tables

Acknowledgements

Much of the material and many of the ideas in this book have arisen from working in collaboration with colleagues on numerous research projects, notably at the Cities Institute; at Maastricht University; and at University of the Arts London. Valuable insights have also been gained through international workshops organised as part of the Regional Studies Mega-Events Research Network and AHRC SmART Cities and Waste Network, as well as conference sessions convened at the AAG, ISA and other international meetings.

This work could not have been realised without partners, too many to single out, but they include arts centres, museums, the London Festival of Architecture, cultural agencies, including the Department for Culture Media and Sport, Arts Council England, Council of Europe, OECD, UNESCO, British Council, and funding bodies, notably the Arts and Humanities Research Council (AHRC), Historic England, and regional governments in London, Toronto, Quebec, Limburg and in the Midlands UK.

Special mention is due to my Research Fellow Dr Ozlem Tasci-Edizel, and the artists Lorraine Leeson, Simon Read, Rebecca Feiner, Arts Researcher Phyllida Shaw, the Hackney Wick Cultural Interest Group, City Fringe Partnership, Three Mills Heritage Trust, Digital Shoreditch and Crouch End Preservation Trust.

1 Introduction

Art is a social product . . . to be seen as historical, situated and produced.

(The Social Production of Art, Wolff 1981, 1)

Twenty-something years ago in *Cultural Planning: An Urban Renaissance?* (2001), I appraised the role and place of culture in the development of cities and urban spaces over time – in classical, medieval, early industrial and late capitalist eras. Manifested over the past 50 years in urban regeneration and latterly placemaking efforts, culture in its various artistic, built, political economic, spatial and symbolic forms was demonstrated to have been used and abused, as an amenity, exceptional and everyday community resource, and increasingly in order to accommodate both consumption and production imaginaries and the growth of the cultural and creative industries worldwide (Wu 2005; Evans 2009c). Much place-based culture, of course, depends on participants – audiences, users, consumers, visitors and modern-day *flaneurs* – in order to be validated, and in a mixed-economy system, valorised. Indeed, as MacCanell opined in 1996, tourism was now the cultural component of globalisation, and although this is less exclusively so today with the advent of digital media, cultural consumption *in situ* still relies on people to go to the place of production, whether or not it is curated and staged (e.g. film studio tour, museum), distributed and experienced every day (e.g. library, local festival, broadcast, cinema), and whether spectacular or sacred, such as pilgrimages and local festivals and national and international events (Bauman 1996). Digital platforms are also increasingly used, as the book reveals, in transporting 'live' places and performance to local places of collective experience, as well as converting 'real' cultural experience to the virtual – from computer games, immersive art and performance to archiving collections.

It has also been 50 years since Henri Lefebvre's thesis on the *Production of Space* (1974) shone a light on the relationship between the urban environment, power and space – local amenity, public and private spaces – and the social context, governmental systems and norms that create and often impose these environments and, importantly, how they are experienced rather than consumed. As Lefebvre observed, you don't *use* a work of art such as a sculpture, but you *experience* it. This sounds obvious today, not least with the identification of the so-called

DOI: 10.4324/9781003216537-1

experience economy (Pine and Gilmore 1998) – as if experience, entertainment and the aestheticisation of consumption were absent beforehand (Paterson 2006) – and a corporate shift from product to brand value, a shift not limited to consumer products but to place itself through the conflated concepts and practice of creative place *branding* (Kavaratzis, Warnaby and Ashworth 2014; Evans 2014a), place-*making* (Markusen and Gadwa 2010) and place*shaping* (Evans 2016). This place-valuation process arises in most of the cultural spaces discussed in this book in both predictable and unexpected places. The latest paradigm of urban economic space is encapsulated in what human geographer Alan Scott terms *cognitive cultural capitalism* (2014). This state, he argues, has supplanted the more instrumental models of post-industrial growth represented by notions of the creative city inhabited by the eponymous Creative Class (Landry 2000; Florida 2002) and further characterised by the nomadic 'creative' and cultural tourist (Richards and Wilson 2006), where 'a core element of the creative fields within these cities consists of the clusters of technology-intensive, service and cultural producers' (Scott 2014, 570). Cultural spaces in their social and creative guises do not tend to feature in these economic metanarratives of the contemporary city, other than through their subsidiary amenity value.

The cultural content flow between the *arts-cultural-creative* production 'chain' and the positive spatial relationship observed between arts and cultural facilities and creative industry clusters (Noonan 2013) provides us with another cultural production and consumption heuristic. It is no accident that many of the *starchitectural* public buildings created in the last 50 years have been cultural – from performing arts and civic centres, museums, galleries and libraries to large-scale sculptures, installations and plazas (Sudjic 2005; Ponzi and Nastasi 2016; Foster 2013), a seemingly counterfactual development in the era of the digital economy and networked society. The desire for collective and place-based cultural exchange has not however diminished – in fact, this has proven to be the corollary to social atomisation – whilst cultural production populates the content industry on which the digital economy relies. A living culture requires space for development, knowledge and skills acquisition, for traditional and new cultural practices, and places that are both accessible and non-exclusive. Where this is not the case, culture can only be partial and ultimately sterile, and the conditions for its social production are therefore fundamental to a society's cultural health.

Lefebvre's triad of space production (1991) – *spatial practice, representations of space* and *representational space* – offers, to a certain extent, a conceptual framework for the consideration of cultural space, but which, as this book argues, also needs to look at culture from the perspective of the user and participant and the relationship between production and consumption. Arguably, this is one of the foundations of Lefebvre's concept, accepting the ambiguity and fluidity of space creation, production and usage, and the non-exclusivity between these modes of perception, conception and lived experience. Culture as an everyday practice, and as a special and heightened activity, is strongly influenced by symbolic and social knowledge, by prior experience and cultural capital. Formal cultural space production is also highly conceived and codified through planning ('technocratic sub-dividers'), functional and affective design norms, whilst cultural content and programming, including the work of artists (Wolff 1981), are also pre-conceived through arts policies, education, tastes and fashions, directly influencing its social production, dissemination and replication. Mainstream culture is also subverted through sub-cultures, avant-garde, radical art and resistance and renewed through culture's essential experimental nature, in addition to traditional arts that exist outside of the prevailing hegemony. Participation in culture of course varies in intensity and engagement, active and passive, but it would be misleading to present the audience member in say a theatre or cinema as a mere observer and *usager* of cultural space, or a museum visitor as necessarily an active participant (Bourdieu and Darbel 1991). Experience is influenced by the social setting within which we live, and the 'space which is presented before an individual is loaded with symbolism and stimuli with a visual essence infused with the social reality from which it exists' (Brown 2020, 3). Furthermore, you cannot play, or learn how to play golf, without access to a golf course, and whilst some culture is self-generated and domestic, collective culture demands spaces and facilities at a quality and scale within which skills and collaboration can be nurtured, cultural artefacts and professional performance can be staged, exchanged and experienced, and here we see the essentially uneven nature of access to cultural experience which has varied both historically and across social and political systems. Notions of cultural rights have in consequence been enshrined in various charters and treaties since Lefebvre's *Rights to the City* of Paris 1968, in which he also included cultural rights (Lefebvre 1968, 1970).

Cultural Rights

Applying this social production concept – blending and juxtaposing ideas of rights to the city (Lefebvre 1996) with notions of cultural amenity and access, and the planning of cultural space – is therefore the driving foundation of this new book. Whilst culture maybe almost benign or secondary in comparison with wider social and economic concerns, it is also the case that cultural rights (CEC 1992; Fisher 1993) and the European Urban Charter (Council of Europe 1992)[1] have emerged as important elements in human rights and in sustainable development principles, with culture now considered to be the *fourth pillar* in sustainability (UCLG 2004; UNESCO 2009; Hawkes 2001) – alongside the social, economic and environmental (Evans 2013). Cultural practices and traditions are also often one of the first victims of totalitarian regimes, war and conflict, as well as vulnerable to the effects of globalisation, commodification and the privatisation of space, whilst hegemonic power also limits greater diversity and resources available to so-called minority and community cultures. This is evident, for instance, in the uneven geographic and financial distribution of resources for cultural facilities and programmes (Evans 2016) and in the response of artists and cultural groups in activism and resistance in the pursuit of cultural and social justice (Lacy 1995).

Social change and demographic shifts have also increased demand for public culture and spaces. Smaller family size, single occupancy living and a growing elderly population have all contributed to this shift, manifested through the demand for more easily accessible social opportunities outside of the home. The workplace has also changed:

> many jobs have been emptied of social and creative content by technology and efficiency measures. More people have more time and at the same time a number of social and creative needs must be satisfied through outlets other than the traditional (home) and workplace – the residential area, the city, public spaces – from the community centre to the main square.
>
> (Gehl 2001, 52)

Global movement and displacement – not just post-industrial tourism – also influence cultural development and cultural space formation and experience through diasporas, ethnic communities and clusters, creating multicultural and intercultural spaces, including museums and new public realms (Bagwell et al. 2012). Conversely, the export of national culture provides another example of *hard branding/soft power* (Evans 2003a) undertaken by cultural and educational institutions through cultural diplomacy programmes (e.g. UK British Council, German *DAAD*, French *Alliance Francais*), as well as corporate expansion seen in major investments in outposts/franchises in regions representing new markets for cultural consumption and university education – for example, in the Middle East, notably Abu Dhabi, Qatar and the UAE (*vis* satellites of Guggenheim, Louvre, Smithsonian) and 'offshore' campuses[2] (UCL and NYU). The production of these cultural spaces is therefore far from benign and less than social in intent. Nonetheless, they represent a long line in boosterist efforts to celebrate and wield 'cultural' power and wealth, in contrast to more organic and engaged forms and places of cultural exchange.

The material used and case studies, which have formed the basis of detailed investigation, have been drawn from intensive and in some cases longitudinal empirical research that I have undertaken – as an individual and group member of teams I have been fortunate to be part of as a result of leading research centres and institutes both in the UK and in the Netherlands – drawn from across the social sciences, humanities and art and design practice, including working with policymakers in the cultural, architecture, planning and environmental spheres, as well as with practicing artists. This is in contrast with much work and writing on urban culture which relies on a superficial and distant relationship with their subject of observation or a technical analysis of enterprises, transactions, buildings and space, and a less-than-engaged relationship with those stakeholders who both produce and experience cultural space in everyday and exceptional ways. These actors (*sic*) are generally reductively represented (or not) through sociological stratification in arts and cultural activity studies, as reflected in secondary and audience/user surveys and datasets (Evans 2016b). The basis of these studies is generally limited, of course, to usage of official culture and spaces, notably the selective high/subsidised arts, gallery and museum venues and heritage sites where socio-economic difference in both frequency and depth of usage (e.g. time spent in cultural venue) is amplified. This narrative has remained largely unchanged (and unchallenged) since

Bourdieu's early work in the museums of continental Europe in the mid-1960s, to the perennial government-sponsored cultural activity surveys today (ibid.), despite the fact that time spent away from home, work and travelling has doubled over this period (nVision 2006). As Gehl again observes: 'Cultural assumption now means ever-shifting networks of taste, activity and enjoyment that are not easily correlated with the usual social categories of race or class, age or gender' (2001, 52). Likewise, the academic and official narratives of the creative industries are typically presented as aggregate groups of anonymous enterprises and firms, divided into now-dated industrial sectors which reflect neither contemporary cultural and creative production processes (in fact omitting much arts and cultural production altogether – Metro-Dynamics 2010) – nor consumption and experience (O'Connor 2007, 2010; Higgs, Cunningham and Bakhshi 2008). As Scott warns us: '"creativity" is also a concept that calls for enormous circumspection' (2014, 566).

This situation has therefore also demanded qualitative research approaches that encompass ethnographic, participant observation, participatory arts and action research, and a longitudinal frame for community and site-based engagement with cultural spaces and their communities. The work drawn upon in this book has also benefited from opportunities afforded through international agencies and collaborative projects, notably between London, UK, and Ontario, Canada (*Creative Spaces*); UNESCO Creative City and ICOMOS heritage initiatives (*Diversity and Development* – UK, Mexico, South Korea); British Council (Eastern China, Quebec); Council of Europe (*Cultural Routes, Intercultural Spaces*); Regional Studies Association (*Mega-Events Research Network*); CABE/CABE Space (*Design Quality*) and the Arts & Humanities Research Council's (AHRC) *Connected Communities* programme. The latter has also supported the development of participatory research through site-based cultural mapping, and art- and design-inspired engagement in collaboration with host communities, cultural clusters and other local stakeholders in cultural space development. In particular, my earlier experience as an arts centre director, musician and arts network director provided deep and lasting insight into the value of arts provision and programmes at neighbourhood level and the importance of cultural space in amenity and community life – a positive relationship I have observed worldwide.

Cultural Spaces

But what/why Cultural Space? As I have argued earlier (Evans 2001), spaces for culture have been an enduring feature of cities and social life generally through the ages, and provided a key distinction that particular cities – and societies – have demonstrated and imposed (Hall 1998). Many of these spaces, facilities and traditions – tangible and intangible – provide important legacies for cultural activity today both physically and symbolically. However, as Scott observed, 'a distinction was frequently made between cities of industry and commerce, on the one hand, and cities of art and culture, on the other' (2014, 569). This is reinforced today through selective creative city and cities of culture and heritage designation and competitions, despite their well-meaning intentions. Cultural spaces are therefore neither universal nor homogenous, notwithstanding the globalisation of cultural organisation and forms – both institutional and popular, and where seemingly homogenised cultural space has been conceived, and its users/inhabitants still negotiate and experience these spaces differently (Lefebvre 1991). So that whilst policy convergence and transference are evident in this field (Peck 2005), and localised models of policy formulation and intervention appear similar – including built forms and themes (*culture and regeneration, city branding, creative city*) – local conditions and variations such as the historical, social and cultural identities, as well as governance and geographies/scales, should be equally considered in order to avoid falling into a reductive trap of universality at the cost of understanding the importance of the particular (Evans 2009c, 1006)

It has also been the so-called cultural turn associated with post-modernity that has emphasised the role and visibility of art and culture in both social change and in a conceptual shift towards meaning, cognition and symbols, not limited to high art. So as Scott goes on to argue:

> Today, this distinction is disappearing in favour of a more syncretic view of cities that is in some degree captured under the rubric of the postfordist city, one of whose declinations is the *creative city*, i.e. a city where production, work, leisure, the arts and the physical milieu exist in varying degrees of mutual harmony.
>
> (ibid.)

This widening of the cultural sphere, Jameson had earlier argued, saw culture as

> coterminous with market society in such a way that the cultural is no longer limited to its earlier, traditional or experimental forms, but it is consumed throughout daily life itself, in shopping, in professional activities . . . in production for the market and in the consumption of those market products . . . social space is now saturated with the image of culture.
>
> (1998, 111; Paterson 2006)

As Craik also puts it:

> [T]he spaces and places in which consumption occurs are as important as the products and services consumed. . . . Consumption occurs within, and is regulated by, purpose-built spaces for consumption characterised by the provisions of consumption-related services, visual consumption and cultural products.
>
> (1997, 125)

At the urban scale, Simmel maintained that consumption also provided a bridge between the communal and individual (1950), and using the example of fashion, mediated the relationship between the self and society. The mediating relationship between production and consumption had been observed earlier by Marx, first arguing that production is also consumption since human energy and other resources and materials are used up in the making, but that consumption was also production, in the case of art and aesthetic production providing a means of enjoying the growth of our imagination and sensibilities (Chanan 1980, 122), and thereby developing aesthetic tastes: 'the object of art, like every other product, creates a public which is sensitive to art and enjoys beauty. Production thus not only creates an object for the subject, but also the subject for the object' (Marx 1973, 88). Whilst production and consumption should not be conflated (Wolff 1981), they are thus complementary, and in terms of cultural spaces, they can, of course, be subject to supply and demand, or structure and agency relationships, but as Bourdieu argued (1993) where the 'economic world is reversed', there needs to be a consideration of the specific conditions (e.g. power, hierarchy/hegemony) within the field of cultural production at a particular time – and place. The role of cultural intermediaries (Bourdieu 1993; Jacob and van Heur 2014) is important here, since they can exert a similar influence over cultural space production to those experts – architects, planners/masterplanners, engineers (social and physical) and other specialists – who conceive public space through representations based on prior knowledge, codification and signs and therefore dictate its access and functions, which can also be subject to change over time and from 'above'.

It would be amiss however to reduce cultural spaces and sites for exchange, experience and participation to inevitable victims of, or collaborators in, commodification and conspicuous consumption. This dominant narrative is persuasive given the evident spaces of consumption (Miles 2010), the seeming collaboration of cultural organisations in urban regeneration (Evans 2005) and place branding (Protherough and Pick 2002, 349), and the conflation of culture in the creative industries as rationales for public investment, preservation and survival. Arguably however, the gradual move towards the democratisation of culture has also enabled wider access and distribution of cultural spaces and programmes and a greater consideration for diversity in cultural production, including social architecture, critical curating and inclusive design (Evans 2018). This has been at the risk of a loss of authenticity, autonomy (e.g. of the artist and participant as agent) however, and arguably innovation, but cultural democracy ultimately should provide an open cultural economy in which these freedoms flourish.

Much cultural engagement is however not consumed or considered to be consumption at all by participants, but is celebratory, communal and participatory – essentially self-generated, from amateur to community arts and arts in education activities, and likewise the spaces that serve as host and sites for everyday arts, crafts, exhibition, performance and memory. The 'everyday' is often posited in binary opposition to the 'special', conceived space of professionals, experts, artists and the super-creative. In this biased sense, everyday experience is the kind of knowledge or *doxa* that Plato had little time for, because it arises from common sense, not the analysis of experts. Common sense knowledge in this view is 'disjointed, episodic, fragmented, and contradictory' (Hall 1996, 431), but since we are all participants, such knowledge is (also) inherently democratic (Duncum 2002, 4). From this perspective, everyday life is mundane (and seemingly unaffected by great events and the extraordinary) and involves the reproduction and maintenance of life, not the production of new ways of thinking and acting (ibid.); however, this both undervalues and omits the value of cultural exchange and engagement (and innovation) which can and does occur at all levels, in formal and informal spaces, through the media, and on the margins, not just (or no longer) in the Agora.

The degree of publicness of cultural spaces is also an indication of their role and contribution to the society within which they are situated:

> There is nothing natural about the public domain. It is a gift of history, and a fairly recent history at that. It is literally a priceless gift. The goods of the public domain cannot be valued by market criteria, but they are no less precious for that. They include welcoming public spaces, free public libraries, subsidized opera . . . the broadcasts of the BBC World Service.
>
> (Marquand 2004, in Worpole 2013, v)

Nonetheless, this axiom has not prevented economists (and self-styled 'cultural economists') from applying market economic valuation methods to public culture, encouraged by government cultural and other agencies, ostensibly to aid in their policy and decision-making.[3]

Book Structure

The cultural spaces and related urban phenomena, which are the subject of this book, thus represent an eclectic but congruent and contemporary range of scenarios of *continuity and change* within which urban culture is both experienced and consumed – and produced – and the social context and forces which have influenced this relationship and the urban landscape. This encompasses both the formal, designated and informal spaces where cultural activity takes place and is practiced – and can be observed (or in some cases, is hidden). This is reflected in opening chapters on institutional and circumscribed cultural spaces and their evolution by movements such as Arts Labs and Centres, and districts where cultural facilities such as museums, galleries, theatres, arts centres and libraries are located; as well as more temporal spaces for events and festivals and their legacies. Events and festivals are a particular phenomenon of fixed and fluid cultural space production that are transformed as they evolve from conception, execution and repetition, producing idiosyncratic and curious legacies – physical, in memory and also in community cultural development.

Sites of and for cultural heritage present a particular *problematique*, not just their interpretation, representation and preservation, but their distinction through a hierarchy of historic selection and designation on the one hand, and lived or everyday and intangible heritage on the other. The next chapter therefore considers the role of heritage in cultural space formation and usage, highlighting the deficit in surveys of participation and attitudes towards cultural heritage, as in the case of engagement in the arts more generally (Evans 2016b). The re-use and adaption of heritage buildings presents a rich scenario for cultural space reproduction, situating these sites of memory and their sense of place as expressed by resident communities and others. Examples and contrasts are presented between world heritage cities, including resident attitudes to living within a heritage site and the radically different treatment of ostensibly similar industrial heritage constructions, but ones located in different geographies and societies – confirming that it is place in the socially produced sense, not the heritage and material content that creates this distinction in lived cultural space. As with the critique of arts and festival spaces, cultural heritage is considered both in its more institutional form, as well as through its relationship to a sense of place and user engagement with local history and sites of memory.

The next section focuses on spaces of cultural production and exhibition, where creative industries cluster and generate place branding opportunities and dynamic zones for innovation and investment, most recently in the digital economy, as well as more traditional sectors such as fashion. The relationship – historic, symbolic, economic and cultural – between early forms of cultural production, emerging industrial districts and newer creative industries clusters is a recurring theme. This is a phenomenon that now encompasses cultural consumption preferences, reflecting, perhaps, Marx's idea of the continuity of production with signification and performativity (Joseph 1998). The coming together of these areas as sites for cultural production and consumption has presented placemakers, investors (venture capital and property) and city-dwellers an opportunity to share in the cultural production scene *in situ* against a backdrop of street art, street markets and social networks, and café culture, clubs and cool clothing outlets. This narrative seemingly plays into the Creative Class concept associated with urban economic growth (Florida 2003; Nichols Clark 2011); however, stronger historic and cultural factors have determined these sites of innovation particularly in their embryonic stage, including cultural production affordances and assemblages which are both more nuanced and complex (Storper and Scott 2009). As Heebels and van Aalst suggest: 'future research should further explore the production side of creative activities and the role of social networks and the quality of place in creation of new ideas in creative sectors' (2020, 361) – a sentiment to which this section seeks to respond through analysis of both case studies and associated literature.

This is therefore explored further in the succeeding chapters on Digital Cultural and Fashion space, representing 'new' and 'old' creative production, respectively, but which despite their global representation, exhibit both place-based preferences and values. The museumification of fashion artefacts and design icons contrasts with its often lowly production sites, but which have converged with creation and consumption both spatially and economically as fast fashion and haute couture embrace. Fashion branding – an extreme financial success story – continues to use its place associations (origins/roots, world fashion cities) and exports this to aspiring cities worldwide through fashion weeks, colleges and new markets. The downside of this productionst paradigm has, however, been waste that has defied physical recycling, pollutes through micro-plastics and non-degradable material, and damages third-world textile production through the export of unwanted clothes, ironically via charity and thrift shops. Here responses through make-do-and-mend movements, sustainable material development and creative re-use offer some light, but fashion production and consumption are both situated and dis-located from place, which has allowed its unfettered and unsustainable growth.

The creative industries are also often fêted as a clean or smoke-stack-free sector (Page 2020) in contrast to manufacturing, none more so than the digital economy (ignoring however the accumulating E-waste, non-biodegradable metals and exponential energy costs of heating and ventilating huge data storage servers, etc.). The notion of creative-digital production districts has also emerged from their antecedents in cultural production (e.g. print and publishing, graphic design) and in those same cultural spaces, rather than as is assumed, as new technology *arrivistes* to the 'thin air economy' (Leadbetter 1999). A chapter on digital-cultural spaces therefore examines this evolution through Big Tech on the one hand, and on the other, the aspirational *Silicon Everywheres*, as digital clusters seek to emulate Silicon Valley, or at least use place branding to establish distinctive production districts, the more successful of which co-locate with existing cultural and creative zones of the city. The arts and entertainment industries have also begun to incorporate virtual visualisation in their creations and productions, with immersive, VR and AI experiences supplementing, extending and replacing live performance and exhibition, raising issues of authenticity, copyright/ownership and autonomy of artists and creative producers. Presenting the digitally enhanced experience in physical space presents particular challenges, as this chapter reveals.

Continuing with the urban cultural realm, where culture is more embedded, literally, in the urban fabric itself, graffiti and street art provide an antidote to the municipal and commercial production and consumption of space, although even here, street art, the art market and place branding increasingly collide as graffiti's *museum with(out) walls* becomes tamed within the gallery, auction house and cultural tour. Graffiti – and its grown-up manifestation, Street Art – experiences a varied but convergent treatment (by academics, politicians, property-owners and the public), as discussed in a chapter through examples of street art in several cities and neighbourhoods, which unsurprisingly coincide in many cases with areas of intense creative production and new visitor destinations.

It is often presumed, of course, that cultural spaces, and in particular venues and facilities for production and consumption, are well known, identified through arts and cultural buildings, centres, festivals and institutions (Evans 2013) – and a pervasive visual culture in public and private space. However, away from city centre, downtown and larger towns which may support theatres, cinemas, museums and galleries, many everyday places lack such provision, or their cultural activity is 'hidden' in community, shared and other spaces, for example, education and faith centres. As Warren and Jones point out, there is a 'a tension between an economic geography emphasis on [cultural and creative] clustering driving success, and the ways in which creativity and communities often operate at the margins, in peripheral places both within and without cities' (2015, 16). This more community-based culture may be overlooked and under-valued, with towns and neighbourhoods being dismissed or criticised as cultural wastelands or deserts. On the other hand, cities and districts seemingly rich with cultural facilities, heritage and superficial activity may in fact be culturally sterile, touristic and staged, lacking the kind of engagement, participation and creativity that may be hidden in more everyday spaces and locales (Evans 2015a).

The concept and practice of cultural mapping (Duxbury 2015) have therefore developed and emerged into the mainstream, which seeks to identify community cultural and heritage assets alongside the formal, offering a more democratic, co-designed/co-created and co-produced approach to supporting a community's cultural resources and spaces. The development and practice of cultural mapping are elucidated with examples from local areas which reveal the importance of access and the role of participatory approaches to identifying community cultural aspirations and the design of cultural spaces. This approach is situated in the field of socially engaged (arts), where it is fitting to consider in the final chapter, collaborations in cultural spaces that take the concept outside of the venue and street-as-canvas and into community and co-designed spatial practices, with examples of artist-led environmental and cultural ecosystem projects in everyday cultural and 'non-cultural' spaces. An important distinction here between cultural intermediaries and intermediation stresses the power shift between professional cultural producer and *prosumer*. This also leads us to consider the role that cultural spaces have in notions of sustainable development and 'the idea that "place" plays in connecting together community, policy and creativity. The key here is not to think of place as a singular, unified scale, somehow more "authentic" because it is closer to "communities". Instead place is relational, cutting across scales and challenging the one-policy-prescription-fits-all model of community + culture = economic success' (Warren and Jones 2015, 16).

Each chapter therefore considers in turn a particular type of cultural space and its production in relation to consumption and engagement and where appropriate, vice versa. This considers spatial scale through individual cultural facilities, heritage sites and through their distribution afforded by events and festivals, and cultural mapping and planning exemplars. Cultural production clusters and quarters present an area-based perspective including their inter-relationship and the growing convergence with consumption practice and the congruent sites of production-consumption in what some would refer to as the experience economy, and others as examples of place branding and gentrification. This is apparent as discussed in chapters on the traditional cultural fields of fashion and in digital content production – both in terms of the immersive-turn in arts production and presentation (more the latter than the former) and in creative-digital production that clearly builds on traditional cultural clusters and spaces, with digital economy and tech-based fashion-making districts defying the *death of distance* (Cairncross 1995).

Cultural production in all senses, as this book argues, reinforces the significance of place and the enduring quality of cultural spaces, and this is not least apparent in sites where social engagement in cultural development is practiced through co-design and co-production. Lefebvre, of course, had presented a radical vision for a city in which users manage urban space for themselves, beyond the control of both the state and capitalism, and by the same token this must encompass cultural space and freedoms. However, the artist is largely absent in models of cultural space production, other than as documenters and describers of representational space (Lefebvre 1991), so their role – and the challenges this raises – in socially engaged arts practice with stakeholders advances the structure-agency dialectical interaction, as the final chapter explores through case studies of artist-based collaborations in and around cultural and heritage spaces.

Notes

1 The European Declaration of Urban Rights included *Culture*, alongside 19 other urban environmental rights: that is, access to and participation in a wide range of cultural and creative activities and pursuits.
2 In 2011, an estimated 100 branch university campuses operated worldwide, over one-third in the Arab region (most in the UAE and Qatar), with the majority opening within the previous decade, nearly 50% of these from the USA followed by the UK (Miller-Idriss and Hanauer 2011).
3 www.gov.uk/guidance/culture-and-heritage-capital-portal

2   A Place for the Arts

The history of cultural venues and facilities – from the grand cathedrals of art, theatre and opera to popular entertainment and municipal and neighbourhood cultural amenities – forms an important part of our sociocultural evolution. That many of these cultural spaces still exist in some form as sites for collective gathering and usage is an indication not just of their longevity but also of their symbolic importance and contribution to a sense of place. Many of the buildings they occupy were clearly built to last and have survived regime and social change. As Schubert maintains, for example: 'The museum remains an exceptionally adaptable cultural construct, both deeply vulnerable to outside interference yet of awesome robustness at the same time' (2000, 153). Their location often anchors urban settlements at many scales – village hall and community centre; town hall and civic square – and they therefore often represent the cultural component of urban growth and new town development (Evans 2001, 2008). In many cities, these single spaces have grown into museum islands and quarters, theatre districts and arts complexes, which have been emulated across the world, extending a cultural circuit and style which, in particular, benefits an international art, architecture and cultural market and associated mileux.

Specific art form spaces (although these have proved to be fluid over time) make up an important strand of cultural history and subject of scholarly study – notably theatres (Southern 1962), opera houses (Beauvert 1995), museums and galleries (Schubert 2000), arts centres (Hutchison and Forrester 1987) – as well as popular entertainment spaces, from fairs, music and dance halls (Crowhurst 1992; Weightman 1992) to cinemas (Chanan 1980). Many of these leisure uses re-occupying the same buildings and cultural spaces over time, as tastes, technology and talent evolve. Their social significance was not confined to pleasure however – Sennett identifies the early urban cosmopolitan with the rising bourgeoisie and the construction of public space (2000), but where the public sphere occurred not just through economic exchange but through 'much more political and social exchange . . . the debate between free citizens in the coffee-houses and the salons, the meetings in theatres and opera houses' (Burgers 1995, 151). A radical and reformist streak has persisted in these cultural spaces, notwithstanding their perceived leisure and recreational purpose and an amenity approach to cultural provision that has seen their independence to an extent blunted by a planning system that prefers

DOI: 10.4324/9781003216537-2

governmental and functional standardisation in the name of distributive equity in cultural provision (Evans 2001, 2008). Since the 1960s, this approach has facilitated (although by no means created) both an arts centre movement and model (Evans 2016), but one that has arguably, with a few exceptions, limited innovation and resistance (Jeffers and Moriarty 2017; Shaw et al. 2006). The evolution of Arts Labs and Arts Centres is therefore a particular focus, including their influence on new cultural spaces.

The place, form and function of culture in the development of cities over time have been covered at length in my earlier book *Cultural Planning* (2001) and some of the issues and phenomena arising will be revisited here, as new cultural facilities continue to emerge as part of urban regeneration, competitive *Creative City* (Evans 2017) and eventful city strategies. Culture and its critiques also suffer from a productionist tendency which is perhaps not surprising given that its proponents and commentators are drawn from this side of the equation and dialogue. Attention to users, consumers and audiences for the arts and culture therefore also warrant consideration in a discussion of cultural space, not just because they/we inhabit, validate, pervert and make real the cultural experience but also because these spaces have ostensibly been developed *for* them/us – if not by/with us in most instances. The idea of design quality is, in particular, considered as an *a priori* method to assess the impact on new cultural facilities from several user perspectives, which not surprisingly produce differing views and experiences. A focus on everyday cultural activity and spaces includes on the one hand the public and commercial amenity provided by libraries and cinemas, and on the other, the notion of intercultural city space. These produced and informal spaces represent different heterotopias which together challenge the concept of immutable culture.

Place for the Arts (Centre)

Over 50 years ago, *A Place for the Arts* was published, combining a survey of arts centres with responses from 25 different countries covering eight world regions and documenting over 170 individual arts centres and arts labs, with supporting architectural plans (Schouvaloff 1970). In the UK, there were almost as many declared *Arts Labs* as there were designated Arts Centres, whilst in Paris by that time, the more dirigiste *Maisons de la Culture* had been established in ten towns and cities. The first *Maison* had been founded in 1935 in Paris, as the headquarters of the Association of Revolutionary Writers and Artists. Both countries embarked on post-War arts centre building programmes inspired respectively by Jennie Lee's *Policy for the Arts* (1965): 'before we arrogantly say that any group of our citizens are not capable of appreciating the best in the arts, let us make absolutely certain that we have put the best within their reach' (in Black 2006, 330); and Malraux's grander *Maisons de la Culture* programme for the French regions: 'a house of culture per department should see the light of day within three years, so that "any sixteen-year-old child, however poor, can have real contact with his national heritage and with the glory of the spirit of humanity"' (Girard 2001, 3). In post-War reconstruction Germany, over 100 theatres were built or rebuilt. Here, a more regionally based political system, with greater devolution and regional independence from the administrative centre/capital, supports higher levels of cultural facility provision, more widely distributed (i.e. regional cities), than in more centrist states such as Britain, France and Greece, where the capital dominates in higher-level provision, such as opera houses, theatres and cultural production. In Federal Germany, unlike London, Paris, and New York where professional theatre is predominantly concentrated, 76 communities maintained a public theatre in 1970–1921 state, 102 municipal theatres (with 88,000 seats), in addition to 27 private theatres, 40 travelling and 18 small theatre companies (Evans 2001).

In the USA, the flagship Lincoln Center for the Performing Arts first opened in New York in 1962, with several additional performance spaces for theatre, opera, dance and music opening between then and 1966, including the Juilliard School and the Library and Museum of the Performing Arts. A further 50 cultural centres were operated or planned by Community Arts Councils across 28 US states at that time (Schouvaloff 1970). A decade before, as part of the post-War cultural reconstruction, the Royal Festival Hall was built to anchor the 1951 Festival of Britain (accompanied by temporary pavilions and domes) and this complex was expanded in subsequent years to include three further arts spaces (Queen Elizabeth Hall, Purcell Room and the Hayward Gallery) hosting music, theatre, art and literature venues (e.g. the Poetry Library), adjoining the new (relocated) National Theatre and British Film Institute (BFI). Less than two miles away, the Barbican Centre was also conceived at this time (although taking longer to realise), on a former bomb site in the City of London. An addition to the brutalist post-War housing estate, this complex of theatres, concert halls, cinemas, library and exhibition spaces plays host to several national live arts companies. Unusually both these major cultural centres are associated with housing, in the case of the South Bank Centre, the Coin Street Builders, a community enterprise/cooperative was formed in the 1970s to campaign for social housing and to redress the declining population and spread of commercial developments in the area. Over 1,000 homes have been built besides the South Bank, all at social-rent, providing affordable housing in a mixed-use area that otherwise was destined to become a mono-use (cultural/visitor) and commercial district.

In Montreal, Canada, a large network of cultural centres has also been developed since the 1980s, directly inspired by the French model. This network, *Accès culture*, has 22 multidisciplinary centres as well as specialist centres, La Chapelle historique du non-Pasteur and the seasonal Théâtre de Verdure, in the 19 boroughs of the city: 'Combining diversity and quality, the Maisons de la Culture offer stands out for its proximity and accessibility. The population can thus take advantage of a programme offered free of charge or at a modest price, in all the boroughs – from theatre to music, through exhibitions, cinema or dance'.[1] In South Africa, a post-Mandela programme of cultural development and the creation of 40 community art centres in black townships and rural areas looked to re-establish indigenous cultural practice and expression effectively banned under the Apartheid regime, as it did in newly independent Zimbabwe in the 1980s with a programme of village-based cultural centres (Evans 2001). Combined with the legacy of museums and galleries[2] and from the 1970s, the opening of country and stately homes to the public, where a 'private heritage became part of the national heritage' (Sassoon 2006, 1367), the second half of the twentieth century has seen the partial move towards access to culture as a democratic right[3] (Evans and Foord 2000). Attendance at live arts events was also significantly higher in Eastern bloc countries prior to the fall of the Berlin wall in 1989. In the decade before, attendance at performing arts events in Hungary, Poland and Romania was over 600 per 1,000 people, compared with 500 in Germany, but only 230 in the Netherlands and Italy, as Sassoon observes: 'The citizens of eastern and central Europe . . . did not have as many consumer goods as those in the West, but they had plenty of culture' (2006, 1250).

Arts Labs

One lesser-known manifestation of this distributive turn has been experimental spaces for cultural expression and activism. The self-styled founder of the Arts Lab movement in the UK was an American, Jim Haynes. In 1969, a dozen or so Arts Labs had been established in the UK, although several did not have full-time premises, or they squatted buildings. By 1970, 150 Arts Labs were operating at some scale (Evans 2001), and arising from a convention of labs held in July 1969, they were organised into eight regions each with a coordinator. The first – and model Arts Lab – was based in Drury Lane, Covent Garden, London, close to the (Royal) Theatre Drury Lane which had operated as a theatre since the seventeenth century. Haynes' definition of an Arts Lab emphasised its multi-purpose and experimental nature, in contrast to single art form venues (IT 1969):

i) A lab is an energy centre where anything can happen depending on the needs of the people running each individual lab and the characteristics of the building.
ii) A lab is a non-institution, we all know what a hospital, police station, theatre and other institutions have in the way of boundaries, but a Lab's boundaries are limitless.
iii) Within each Lab, the space should be used in a loose, fluid, multi-purpose way; a theatre can be a restaurant, a gallery, a bedroom, a studio, etc.

The majority of Labs were not based in existing arts venues, but took place in redundant or adapted schools, pubs, youth and community centres, warehouses, potteries, police stations and even a monastery. The original Arts Lab was a venue for alternative art exhibitions and performances, and although it operated for only a couple of years (its closure occurred soon after it was reported that it had fallen into rental arrears and that the landlord had begun to threaten it with eviction), the facilities were extensive, including a gallery space, a cinema, a theatre and a film workshop. Several of these areas had their own dedicated directors. The cinema was noted for running films all through the night, and passing visitors would sometimes sleep there to save money on hotel facilities. A printed newsletter was produced in connection with the Arts Lab from 1968, initially by British social reformer Nicholas Albery. Among the famous artists and performers whose work appeared at the Arts Lab were David Bowie, John Lennon, Yoko Ono, and the 18-year-old American filmmaker Wheeler Winston Dixon. Following the closure of the Arts Lab, a London New Arts Laboratory was formed by some of its founding members together with new associates. Programmes at Labs reflected local interests and talent, but as well as a range of music, performance, visual and film art, experimentation and multi-media activity; this was supplemented with workshops for these and for printing, publishing and poster-making. Arts Labs were thus 'part of a counterculture that led to the ethos of immediacy of decision-making and urgency of action' (Wilson, in Curtis 2020, 2). Some operated as communes or what would also be referred today as live-work spaces, with some resonance with artist communes (and artist residency programmes) that established across the USA and Europe, although they were largely private rather than open public arts spaces. Operating largely outside of the arts funding system (which did not know what to make of, or how or treat, these unruly upstarts), their financial precariousness and lack of tenure meant that many did not last long, but groups moved on and in a number of cases established community arts centres, in a few cases on the sites of the Labs, or in collaboration with local authorities and other supporters, developed new venues and companies.

Arts Centres

Coinciding with the growth of arts centres as a socio-cultural project, Arts Labs provided both inspiration and distinction from mainstream cultural facilities and provision and stressed the importance of place and of networks. The combination of art forms, cross-arts and media development, amateur and professional, education and training, and ties to communities defined many new arts centres, although the balance between art form priorities/specialisation, professionalisation and experimentation/new work varies, as does their spatial form. Some UK arts centres pre-dated the Arts Lab experiment, for example, the MAC[4] in Birmingham, Arnolfini and Watershed in Bristol (Presence 2019), the ICA (Muir and Massey 2014) and Centre 42[5] in London, but multi-art and media programming and agitprop activities were increasingly evident in these existing spaces and Arts Lab activists continued to work through touring and writing. Key playrights cutting their teeth in arts-centre alternative theatres include writers Mike Stott, Henry Livings, Michael Stevens, Wolf Mankowitz and Edward Bond. Tom Stoppard developed several of his key one-act plays at the Almost Free Theatre (AFT), including *After Magritte, Dogg's Hamlet* and *Cahoot's Macbeth*. His highly successful *Dirty Linen* and *New-Found-Land* went on to transfer from the Almost Free to run for four-and-a-half years in the West End. The AFT was an alternative and fringe theatre set up by American actor and social activist Ed Berman in 1971, who founded the Inter-Action Arts Centre (run as an urban kibbutz-style co-operative) located in a derelict bomb site in north London (and occupying Cedric Price's only built *Fun Palace* – of sorts). Audiences paid what they could afford, but at least one penny. It also pioneered the lunchtime performance, which bought in a whole new audience. The theatre staged seasons, including the first season of gay plays in Britain, the first women's season, a Jewish season, an anti-nuclear season and a season to mark the 1976 American Bicentennial. Inter-Action also created the Professor Dogge's Troupe, a touring children's theatre company that played to many thousand young people each summer, and the largest youth dance and drama programme in the UK, the Weekend Arts College (WAC), started by two local secondary school dance and drama teachers.

Many arts organisations would describe themselves as having a social as well as an artistic purpose. What distinguishes arts centres is that their artistic and social functions are seen as interdependent, as mutually reinforcing. Professional artists and the public are valued equally (Shaw et al. 2006):

> The ivory tower model of elite artists and arts organisations to cater for a narrow segment of society has crumbled. Forward-looking arts organisations now focus on the coalescence of artistic excellence and social impact, the broader health of their communities and offering public spaces that truly enrich the high streets they inhabit.
>
> (Connor and Barlow 2023, 1)

The arts centre sees itself as having a creative relationship with both, as a provider of opportunities for individual development (artistic, professional, educational and personal), and in many cases, their organisational culture and their programming reflect this. The opportunities that arts centres choose to provide are therefore the result of a combination of factors: their origins, location and facilities, their governance and leadership, as reflected in the history of the movement. Users' views on the benefits of arts centres confirm the advantages of variety/diversity, social and community focus, although this varies between urban and non-urban spaces (Table 2.1).

Most arts centres were established as the result of action by a local resident or group of residents and/or artists; action by an arts or community organisation to establish or improve a facility; the development of new school, college or university facilities; a local authority seeking to improve local provision or, more recently, to 'regenerate' an area (Evans 2005). The geographic location of an arts centre, in relation to other provision, has a direct bearing on its culture and its programme (used

Table 2.1 What Audiences Like About Arts Centres (Shaw et al. 2006; Evans 2016b)

Attribute	Non-Metropolitan	Metropolitan
Type of provision	Diversity in a single venue Provision under-one-roof Meets a variety of interests	Centres for ideas, discussion and debate Eclectic, adventurous, innovative, risk-taking Non-mainstream
Ethos/ambience	Something for everyone Community focused Local Range of audience Friendly Mix of activities in social environment	Opens your eyes Fresh ideas Build trust in venue to take audience on journey of discovery Involvement in the process of creating art Can feel unwelcoming/elitist

here to mean everything that an arts centre offers, from performances, screening, readings and exhibitions to classes, workshops, to advice for artists, studios and rehearsal rooms). In many cases, arts centres were opened to fill what their founders saw as a gap in provision. As Stark documented of their rise and proliferation:

> This phenomenal growth is in no sense the result of central, regional or local planning by any one agency, least of all the Arts Council [observing that] their unplanned status meant that there was never enough food for them on the table; they are architectural opportunists – over 80% of arts centres were housed in second-hand buildings, from churches, drill halls to town halls, over 50% of urban centres were in buildings over 100 years old; they are economic and efficient – multi-use/purpose, weekday/end/evening opening; and they are masters of disguise – in terms of their programme, purpose, attracting a wide mix of funding, in addition to 'arts' subsidy.
>
> (1984, 126)

Whilst cultural re-use of civic and industrial buildings can capture and conserve both symbolic and intangible heritage (Evans 2022), the exterior of a building is also vital to the way it is 'read' or perceived, as Shaw et al. report: 'Arts centres in converted buildings can be misread or, worse, invisible' (2006, 9), and despite the fact that community arts centres in particular can be embedded in their neighbourhoods and localities, their institutional past, impervious design and potentially negative associations and memories can act as barriers to engagement and entry. This observation provides some important clues to the design and location of accessible and welcoming cultural spaces, both adapted and newly built, as much as their operation and programming.

In terms of usage, non-metropolitan attenders tend to attend more frequently, reflecting both the diversity on offer at their local venue and the fact that there is little or no comparable provision nearby. Audiences for metropolitan arts centres generally have more choice and attend these centres less frequently than their non-metropolitan counterparts (Table 2.2). This access-spatial relationship is borne out in studies of local participation in a range of arts and sports activities – where the better supplied sports facilities (e.g. halls, pitches, pools) demonstrate more frequent usage, but arts centres – less distributed and numerous – reach a wider participation across the population as a whole, but who participate less frequently (Evans 2001, 125).

Table 2.2 Frequency of Attendance at Arts Centres (Shaw et al. 2006; Evans 2016b)

Frequency of Attendance	All	Non-Metropolitan	Metropolitan
Once a week	15%	15%	4%
Every couple of weeks	11%	13%	
Every one to three months	53%	53%	8%
Once a year	12%	11%	19%
Less than once a year	8%	9%	8%

Arts centres also provide a diverse range of activities and facilities (e.g. studios, classes/workshops) although some may be focused towards particular arts practice – drama, dance, visual arts, media and ethnic arts, for example Yaa Asantewaa (Carnival Arts) and Tara Arts (South Asian) in London. This is reflected in audience take-up of more than one art form with a significant proportion engaging with three or more art form activities (Table 2.3).

Serving a primarily local and sub-regional catchment provides an indication of the benefits of both access/proximity and continuity of provision, which are important for sustained participation and cultural development. For example, in their study of demographic indicators of cultural consumption, Brook, Boyle and Flowerdew and found that over a five-year period from the opening of a new arts venue, the number of households attending arts activities from the catchment area increased by 1,000–2,700 and as the authors suggest: 'the local availability of a venue can broaden the range of people attending' (2010, 25). Another example of the influence of a change in supply, between 2005 and 2009 attendance at museums and galleries in the north-west region, increased from 40% to 47% attributed to the build-up and effect of Liverpool's *European Capital of Culture* in 2008; although attendance slipped back to 45% in 2009 as the level of provision and novelty declined (DCMS 2010). Access to a range of arts and cultural activities and facilities within a neighbourhood locality would therefore look to provision at the same scale as, say, junior schools, not limited to town centres, but accessible at a similar level, whether formal community arts centres, or multipurpose facilities, or where professional programmed and supported cultural activities are delivered via community centres and other venues (Evans 2008). Arts centres in various forms and scales thus represent one of the few distributed cultural facilities that combine more than one art form/cultural activity and space (MacKeith 1996). Where established they can fill a particular gap in local provision, although receiving less attention in arts attendance surveys which tend to be art form/discipline-based.

In the UK, since 1994, new, refurbished and converted buildings, financed in large part by the National Lottery (not just in the arts, but in community, sports and heritage), did however increase competition for artists and audiences, and some of the more established arts centres had to review their artistic direction as a result. This is also the case in cities around the world indulging in culture-based

Table 2.3 Percentage of Audience Attending Different Numbers of Art Forms (Shaw et al. 2006)

Number of Art Forms Attended	All	Non-Metropolitan	Metropolitan
0 (social users only)	3%	4%	1%
1–2	30%	30%	25%
3–5	38%	35%	42%
6–10	22%	23%	23%
11+	6%	7%	9%

regeneration of city centre sites and major upgrades of arts venues, galleries and cultural visitor attractions, driven in part by international competition and the need to attract touring companies and exhibitions. With new streams of capital investment, there are also new arts centres that have failed, or had a faltering starts, including those theme-based facilities that struggled with the museumification of popular culture, or had vague themes and purpose, for example, Sheffield's Pop Music Centre, Manchester's *Urbis* and West Bromwich's *Public*. The *Public* digital media centre had been forced into administration as cost over-runs (from £40 million to £62 million) undermined this overly complex facility, with no clear artistic function. When the digital gallery had to be closed as the computerised exhibits failed, the Arts Council withdrew its outstanding funding less than a year later, leaving the centre's future uncertain. The local perspective on this cultural venue encapsulates the difficulty faced in developing new *edutainment*-based arts facilities in an area poorly served by 'everyday' amenities. But clearly the local community had not been engaged or considered in such a top-down, *starchitecture*-conceived venture:

> [T]he public is a complete waste of money! Sandwell and Black Country needs better schools, cinemas, theatres, swimming baths, ice rinks etc., not a £52 million white elephant! I have 2 children aged 12, 9 they rather spend £20 at the cinema watching the latest pixar film which has more artistic merit than all of you clowns who run, and said yes to the doomed project in the first place.
> *(Building Design*, 13 June 2008)

Despite their longevity and, to a large extent, their universality across socio-political systems *(Maisons* and *Casas de la Cultura*, and Eastern European *Palaces of Culture*), the term Arts Centre, whilst generic but variable in practice, has also become a municipal concept and less enticing in the context of a wider cultural and entertainment provision. From the 1980s, the community arts movement and associated socio-cultural rationales also fell foul of *dirigiste* arts policy – led by economic importance (Myerscough 1988) and urban renaissance imperatives (Evans 2001) – and consequent funding regimes, as well as the associated liberalisation of leisure and consumption spaces. With hindsight, whilst many of these centres developed unique arts production and acted as a reflection of their times, particularly in attracting young artists and playwrights, they generally did not achieve or sustain a cultural democratic role (staffed by the well-educated, often pursuing an 'alternative' culture/lifestyle), and few if any, could claim a place in the cultural democratic process:

> arts centres have slipped from a dominant position in the cultural chain – they've lost part of their role as value-givers within the subsidised world. . . .
> The vortex of commercial popular culture and the consumer, commodity and technological revolutions has taken so much of what the arts centre world values most – smallness, locality, the love and sense of community.
> (Wallace 1993, 2)

However, to a certain extent, mainstream subsidised arts organisations had from the 1980s adopted education and outreach programmes – strongly led by funder policy and arts centre precedents, such that today few cultural venues would not integrate education and community-based activities as part of their main provision, although their cultural space production and experience still differ symbolically and functionally from the arts centre model:

> the inherent features of arts centres – their multidisciplinary context and the community, civic and artistic role they play – can make it difficult to pin down their identities, and often lead them to being underrepresented in the media and public dialogue around the arts.
>
> (Connor and Barlow 2023, 1)

Today, new and re-invented centres have nonetheless eschewed the arts centre title, using more evocative or placed-based names, for example *The Lowry* in Salford, the *Phoenix* in Leicester (the Leicester Arts Centre Ltd), *artsdepot* in north London (formerly the Old Bull Arts Centre), and the *Arc*, Reading ('a cultural destination') and the *Arc*, Stockton (replacing the Dovecot Arts Centre and cinema), including those referencing their host city's industrial pasts. 'Art centres, galleries and museums have also embraced industrial architecture and associations – art is now production and work – *Powerhouse, Gasworks, Leadmill, Printworks, Foundry* and *Factory* are all arts centre names, whether occupying former industrial buildings or not' (Evans 2003a, 420), but their essential distinction from mainstream and single-form buildings persists, at their best by creating different entry points for the public to different art forms and genres (as both users/audiences and active participants), a creative relationship with artists and the public, and a social dimension and space dedicated to social interaction (Shaw et al. 2006). This combination of 'arts-as-amenity' (*Every Town Should Have One*, Lane 1978) and cultural production arising from individual creative and local initiatives – often place-based and extending cultural practice to new audiences/users and new places, especially outside of capital and major cities – has however generally not been co-designed or co-produced in terms of local cultural aspirations, needs or participation (other than on a post-hoc basis). In an important sense, this represents a particular point on the continuum between the autonomy of the arts (artists/companies, directors/curators) and cultural democracy, a tension that exists in public culture – and in cultural policy and resourcing – and therefore in notions of cultural planning (Evans 2001, 2008).

New Wine in Old Bottles/Shock of the New

These more planned and spatially determined approaches to neighbourhood-level arts development are in extreme contrast with flagship public and private/commercial cultural spaces. On the one hand, the tradition of building on existing cultural and other industrial buildings continues, which is not surprising as twentieth-century de-industrialisation also accumulates in terms of redundant buildings and spaces – and host communities who are the subject of structural change from 'above' (Islam and Iversen 2018). These tend to be large scale, although earlier re-use of industrial and civic buildings has been a feature of arts centre occupation of smaller spaces. Examples of cultural re-use include large post-industrial complexes converted into creative industries and entertainment centres in Maastricht, the Netherlands, such as the Sphinx former ceramics factory, and in neighbouring Belgium, Genk's C-Mine (Figure 2.1), whilst the redevelopment of contrasting heritage buildings has taken place on the back of major placemaking, event and regeneration schemes such as in *Euroméditerranée* Marseilles, *European City of Culture* in 2013, notably the flagship *Muceum* fort building and *La Friche* cultural centre.

The Sphinx factory building comprises 100,000 m² of former industrial floorspace in an area close the city centre and now billed as a cultural quarter, which includes the Architecture Centre: Bureau Europa, one art house and one commercial cinema, shops, workshops, a student hotel and requisite cafes and restaurants. With reference to its past, a 120-m-long *Sphinxpassage* uses 30,000 ceramic tiles tell the story of the former Sphinx factories established in the 1830s. Maastricht, an historic, tourist and university town in the catholic-south of the Netherlands, had been shortlisted to host the 2018 European Capital Culture, but lost out to Leewaurden in the north; however, its event and culture-led regeneration programme has continued (Evans 2013b), buoyed by its transborder *Euregional* location and status. This contrasts with Genk in neighbouring Limburg, (Flemish) Belgium. Genk had

Figure 2.1 Sphinx, Maastricht and C-Mine, Genk
Source: Photographs by the author.

been a coal-mining area from 1900 until the 1970s with the last mines closed in the 1980s (and prior to its new industrial landscape, had been popular with artists and its then natural landscape). The mines had attracted migrant workers from Europe, Turkey and beyond, meaning that, today, there are over 100 different ethnic groups living in the city. As mining declined, alternative industry was sought, and in the 1960s, Ford Motor Co. opened a major factory, with its favourable connectivity (canal water transport, one hour drive from Ford's HQ in Cologne) and employing 5,000 people. However, this factory was short lived, closing in 2014 due to over-capacity and transfer of car assembly to Valencia, Spain. The heavily contaminated 140-acre site became redundant and derelict.

In 2000, the city was already pursuing new, 'post-industrial' strategies, and based on models elsewhere (Evans 2005), the idea to develop a creative hub in the buildings of the old coal mine of Winterslag emerged, and in 2001, the city bought the site, and christened the venture *C-mine* in 2005. The centre has four elements: education, creative economy, creative and artistic creation, and presentation. With a university college specialised in various arts subjects, an incubator for young entrepreneurs, a cultural centre, a design centre, a cinema, the C-mine has so far created 330 jobs in 42 companies and organisations, including around 200 jobs in the creative sector. These companies together produce games, apps, websites, sets for television/film, light shows, product design and stage productions, and the centre hosts cultural events, including the international art exhibition, *Manifesta*.

The contrast between the historic international, university town with the advantages of tourists and an established arts and cultural infrastructure, and Genk's post-industrial, multicultural legacy is apparent, but, in many respects, the C-Mine and new arts facilities in the town (e.g. art galleries) are creating a more cultural space, less reliant upon the market and cultural consumer than its more bourgeois Maastricht neighbour.

The Museum of the Civilizations of Europe and the Mediterranean (*Mucem*) organisation is spread across three sites – along the sea front, at the entrance to the Old Port, the new J4 building (designed by Rudy Ricciotti and Roland Carta), and the Fort Saint-Jean (Figure 2.2), a fully restored historical monument linked by two footbridges host major exhibitions and cultural programmes, whilst in the town, in the poorer Belle de Mai district, the Centre for Conservation and Resources houses the museum's collections. The *Mucem* opened in Marseille in June 2013 to mark the city's European Capital of Culture event, representing the flagship of the *Euroméditerranée* regeneration project, at a cost of €191 million – and soon became one of the most visited museums in the region.

La Friche in contrast is located in one of the poorer districts of an already-impoverished city, with embedded socio-spatial segregation and socio-economic inequalities (Gripsiou and Bergouignan 2021). This former tobacco factory had been converted in a large arts centre – in the capital Paris, abandoned urban plots or *friches* have been transformed into cultural hubs with a variety of uses: exhibition spaces, concert halls, workspaces, premises for voluntary organisations and venues for club nights. First occupied by the *Systeme Friche Theatre* (SFT) in

Figure 2.2 La Friche and Muceum, Marseilles
Source: Photographs by the author.

1992 for a novel rehearsal and production space, other companies joined them, including the *Aides aux Musiques* and a 'host of agents are co-present in the same space-time frame involved in a process of artistic creation' (Bordage 2002, 17). Today, the *Friche la Belle De Mai*, giving its full title and location, is both a workspace for 70 organisations including 400 artists and other creatives and for cultural events – 600 public art events per annum, from youth workshops to large-scale festivals. With over 450,000 visitors a year, the Friche is a multifaceted civic space comprising a sports area, restaurant, five concert spaces, shared gardens, a bookshop, a crèche, 2400 m^2 of exhibition space built for the European Capital of Culture, 8,000 m^2 roof terrace, and a training centre with several skateboard parks abutting the railway track above. This arts centre therefore pre-dated the city event but, like other ventures, benefited from further funding to upgrade its facilities.

In both cases a tale of two cities/cultural festivals perhaps – in Marseilles, one largely touristic, symbolic and costly, situated in a high profile and accessible part of the city-by-the-sea, the other, an intensively and democratically used/organised, occupied and flexible space, re-purposing industrial heritage in an area off the visitor map where 'there were no cabs to be found outside the Friche. The neighbourhood of *Belle de Mai* is poor, not prime for cabbies' (Kornblum 2022, 87).

Plug-In Cultural Spaces

Dedicated cultural spaces need not be permanent buildings or planned events, but can also be open and mobile combining the benefits of a recognisable cultural facility with flexible location through a form of portable architecture (Handa 1998). Two examples are Aldo Rossi's *Teatro del Mondo* and Tadao Ando's *Kara-za* theatre. Rossi's albeit temporary project was built in 1979, for the Theatre and Architecture sections of the 1980 Venice Biennale.[6] With a height of approximately 25 m, comprised of a cube topped by an octagon, it was made of steel poles and a wood skin. Anchored at the Punta della Dogana, this ephemeral floating theatre then sailed across the Adriatic to what was then Yugoslavia, visiting Dubrovnik among other ports that were part of the former Venetian colonies, after which it was dismantled. Through his many drawings for the theatre, Rossi analysed the Venetian identity, its physical, geographic, architectural and mythical reality. In it, many references converge: primitive, proto-renaissance Florentine kiosks, renaissance theatres, Elizabethan theatres, lighthouse architecture and especially eighteenth-century Venetian architecture, known for its floating structures built for Carnival.

A more functional example is Japanese architect Tadao Ando's *Kara-za* mobile theatre (Ando and Futagawa 1990). Temporary buildings are an integral part of Japanese culture. The essence of ascetic Shintoist philosophy which pervades Japanese life is that it is not the permanent elements of life that are revered but the spiritual renewal represented in the periodic cycles of death and rebirth, destruction and recreation (Kronenburg 2003). Constructed from scaffolding with a simple assembly procedure, the original theatre was built in 15 days in 1987. Ando designed the theatre to be portable, with a vast majority of its structural elements made from locally sourced standard components. It was first used as a temporary theatre in Asakusa and then at a drama festival in the Kansai district before moving to other festival sites. Another of Ando's design has been constructed as part of the MPavilion 10 festival which opened in Melbourne in the summer of 2023, an annual five-month design festival of free events in Melbourne's Queen Victoria Gardens. This will be the tenth MPavilion event, which has grown to become one of Australia's most-visited festivals, attracting more than 350,000 people in 2022. The first nine MPavilions have welcomed more than 900,000 visitors and hosted over 3,500 free events since its establishment in 2014. At the end of each MPavilion season, the host Naomi Milgrom Foundation gifts the pavilion to the people of Victoria and relocates it to a new, permanent public home in the community. To quote: 'For this intervention, conceived as a new meeting place within Melbourne's Arts Precinct, Ando's MPavilion design encapsulates his desire to create a memorable structure that responds directly to the Gardens. It strives for spatial purity, employing the geometry of circles and squares to create a space in harmony with nature'.[7] However, constructed from concrete and topped with an aluminium disc, the notion of then relocating this structure to another location belies its place-specific design and sustainability. These more aesthetic examples were also pre-dated by the more down to earth concepts of the *Plug-In City* and *Fun Palace* designed by Peter Cook/Archigram and Cedric Price, respectively. Both drew on thinking around modularism and a vision of temporally adaptive and anticipatory space. Here modular units could be arranged and re-arranged, buildings could expand and contract, time and time again and where real-time data on human activity could control and modify the spatial form (Mathews 2005). Prescient in many ways of the immersive and data-driven experiences now emerging from AI/VR exhibitions and displays, as discussed further in Chapter 6.

Shock of the New

Whilst re-use has some appeal and can meet heritage conservation imperatives, new arts and museum buildings and their satellites/franchises, as well as major upgrades, for example MoMa New York, Tate Modern London and the Louvre Paris, have also proliferated. This is in part a reflection of a political preference for the 'shock of the new' over the old (Hughes 1991) and in part the imperative for offering *edutainment* and engagement (Pine and Gilmore 1998) offered by new technology and digital experiences, as well as pragmatically increasing capacity for their accumulating and largely unseen collections. In terms of their replication however, it has been the globalisation and *hard-branding* (Evans 2003a) of western culture (*vis* Guggenheim, Smithsonian, Louvre, Ivy League universities) and its adoption in, and export to, developing countries, notably the Middle East, with city/states seeking to develop their post-oil producer economies, such as Dubai and Abu Dhabi (e.g. Saadiyat Island) and Qatar (Riding 2007; Gluckman 2007). This is also seen in post-industrial China with vast estates of former manufacturing complexes that dwarf those of the west (Chen, Judd and Hawken 2015); and in south-east Asia, for example, Singapore's *Esplanade on the Bay* theatre complex and spectacular bio-park (Evans 2003a). This trend has widened and bankrolled a touring circuit of art and museum collections and performing companies – and of course, architects/masterplanners, curators and cultural intermediaries – the prime conceivers of cultural space (Stanek and Schmid 2014). As Lefebvre put it, these:

> spaces of leisure exemplify the reproduction of capitalism through the production of space: they result from the 'second circuit of capital' in real-estate investment that compensates for the tendential fall of the average rate of profit in the primary circuit of capital, related to manufacturing.
>
> (2014, 160)

They are, in this respect, sites of the reproduction of the bourgeois cultural hegemony over everyday life (Stanek 2014; Lefebvre 1984).

Alongside the sovereign wealth-funded arts and entertainment complexes and competitive city hall investment in downtown cultural facilities, the age-old phenomenon of the individual patron also continues to create new extravagant cultural experience venues. Whilst state and corporate sponsors regularly put their name to venues (e.g. Disney, Guggenheim, Thyssen), property owners have a comparatively free reign. An example is the MONA (Museum of Old and New Art, or 'O'), in Hobart, Tasmania (Figure 2.3). Here, David Walsh, a local boy-made-good, had acquired a bankrupt winery at Moorilla in the 1990s. Billionaire Walsh had made

Figure 2.3 Museum of Old and New Art (MONA), Tasmania
Source: Photographs by the author.

his fortune as a professional gambler who literally beat the odds in casinos. On its hilltop site, MONA evokes the temples of ancient Greece (visitors approach by boat). At a cost of $80 million, the museum is an engineering feat, three levels completely underground, carved into the sandstone cliffs of the Berriedale peninsula.

Described by Walsh as an un-museum: from bodily functions to bestiality, euthanasia to evolution, death to deviance, addiction to atheism, MONA will explore them all. Walsh doesn't call it his 'adult Disneyland' for nothing. MONA's site notes take a suitably tongue-in-cheek approach to the question, what is this space?

- David's fitness marker: think of him as a peacock and his paintings as his feathers
- Sea World with cocktails and a few works of quite-good-but-not-amazing Australian modernism
- Somewhere people can come to say 'not sure about the art but the architecture is amazing'
- Nice spot to listen to some live music
- A really elaborate marketing stunt

This cultural facility and 'experience' are unapologetically non-institutional, but both eccentric and the ultimate vanity project – hardly unique in this sense, but a case of new money versus old. Not a product of planning, expressed need, or resident or artist aspirations – other than his own – but its novelty may persist and the 'O' undoubtedly puts this city on a cultural map of sorts, no less so than Guggenheim Bilbao or numerous *starchitect* art museums and performance venues where form dominates cultural function.

New Town – New Theatre

Where planning does attempt to meet an evident gap in cultural provision, there can still remain questions around what *kind* of space may meet future creative needs. In the context of growth, physical and social, this can be illustrated in the case of Milton Keynes (MK) – located in the South Midlands, the iconic planned, post-war new town – in the development of its civic theatre. Whilst no regional arts planning exercise was undertaken for this new town, the potential for a large theatre was highlighted in the original development blueprint in the 1970s. In 1985, the MK Development Corporation reported that the creation of a live performance space would be highly desirable and, following a successful bid for National Lottery funding, an award of £20 million was made towards the £30 million cost of a theatre and gallery. In 1999, the theatre opened: in the words of the Council, in addition to bringing a variety of performances to the city, Milton Keynes Theatre provides a focus for the city's already-thriving cultural life (in Evans 2008, 77).

From an 'insiders' perspective, however, this traditional theatre is felt to lack a certain spirit. In response to the question 'what would you do to make MK a place where arts were a contemporary and necessary experience?' the renown theatre director Sir Peter Hall replied: '[B]uild a smaller theatre for a start. The present theatre is a dehumanising space. It's well attended because, presumably, there is nothing else that gives you the beginnings of that kind of experience, but it's not a congenial theatre' (Hall and Hall 2006, 237). His namesake, the academic planner added, seemingly without irony[8]:

> I think MK is difficult precisely because it is so completely new. MK central
> is the most totally created, planned space that we have in this country . . . but
> I think the problem with MK is that it has been too successful. So it does not
> have any derelict spaces.
>
> (ibid.)

The distinction between (artistic) content, the flagship facility and the importance of 'place' – cultural and symbolic – is apparent from these observations. The idea that building a new theatre is necessarily the *right* type of provision, or the complete answer to local cultural provision is obviously questionable (Evans 2005), particularly given the realities of funding a venue reliant on touring shows and with no in-house production resource. A 'thriving cultural life' may not be the impression that either residents or visitors would have of this new town. Flexibility over cultural facility location and future needs suggests that responsive design and informal spaces, as well as dedicated production and more participatory-oriented facilities, are required in order to accommodate local needs over the lifecycle (Evans 2008, 90). This could in turn offer residents: 'the freedom to decide for themselves how they want to use each part, each space' – as Hertzberger goes on to suggest: 'the measure of success is the way that spaces are used, the diversity of activities which they attract, and the opportunities they provide for creative reinterpretation' (1991, 170).

Everyday Cultural Spaces

To a certain extent, the twentieth century arts centre, as already discussed, has provided one element in everyday culture, but only where programming, access and representation reflected (and reached) local society. The perennial barriers to participation in arts and cultural activity are not only 'access' – location/ transport, cost/entry price – but also relevance: 'subjects I am interested in' and the quality of cultural activity and events on offer – in short, community and more vernacular culture that reflects the experience and interest of local audiences and participants (Evans 2009d). In cultural (space) terms, the everyday infers a frequency and therefore proximity of engagement and experience, the place of cultural activity in everyday life. The absence of provision on the one hand can limit engagement, as in the USA, where 35% of Americans 'did not agree that there were plenty of opportunities to take part in arts and cultural activities in their neighbourhood or community' (NEA 2019, 3). On the other hand, whilst prior experience (and access) is associated with increased levels of arts engagement generally (using education/parental education as a proxy) and lower income/social class and poorer health is correlated with lower attendance at arts events, these factors are not associated with participation in arts activities or creative groups – or even being interested as a 'non-attendee' (Bone et al. 2021). The everyday encompasses the self-generated 'private' realms – in the home, community groups, faith centres, schools, through traditions, cyclical celebrations and in modern urban planning terms, amenities for enjoyment, participation and collective engagement.[9] Most creative practice and 'making', such as crafts, amateur and youth art, are still undertaken in vernacular (or everyday) settings, including ethnic and community culture, and in interstitial spaces, from the skateboarders and young graffiti artists outside the concourses and undercrofts of the arts complexes of the South Bank,[10] London and MACBA, Barcelona, to the more than two million ballroom dancers that meet regularly in local halls and clubs, even before the advent of TV's *Strictly Come Dancing* – and presumably one reason for its audience success, attracting six to nine million viewers each week (Evans 2009d). Moreover, in the extreme of fringe cultural display and exchange, locations are also more often to be found on the edge of the city, such as raves in warehouses or fields, which commonly attract audiences from a 50-mile radius, and weekly community markets – selling crafts, antiques, food, clothes and household goods – as in ethnic quarters and in second- and third-world cities, under the shadow of motorway flyovers and football stadia. Dance, music and entertainment acts intermingle with these markets, which regularly draw participants from a wide area of the city and surrounding region (Evans 2009d).

Significant cultural engagement is apparent therefore in so-called amateur and voluntary arts activity (more active than passive attendance at professional arts events). In the UK, an annual survey reveals the scale of engagement (Dodd 2008), which included 50,000 organised groups represented by nearly six million members and 3.5 million further volunteers taking part in over 700,000 events attended by 158 million during the year (Evans 2009d, 27). Whilst amateur dramatics and music are the most popular, 'multi-art', including ethnic and new art forms, makes up the largest and most popular range of activities undertaken (Table 2.4) as in the case of arts centres (as mentioned earlier).

Table 2.4 Amateur and Voluntary Arts Participation

Art Form	Groups (000s)	Members (000s)	Extras/Helpers (000s)	Performance/ Events	Attendance (000s)
Craft	840	28	13	3,000	924
Dance	3,040	128	12	57,000	10,906
Festivals	940	328	395	12,000	3,481
Literature	760	17	11	4,000	191
Media	820	62	12	21,000	1,563
Music	11,220	1,642	643	160,000	39,325
Theatre	5,380	1,113	687	92,000	21,166
Visual Arts	1,810	265	52	8,000	1,289
Multi-art	24,330	2,339	1,692	353,000	79,789
Total	**49,140**	**5,922**	**3,517**	**710,000**	**158,634**

Libraries

Some of the more quotidian cultural spaces in urban life are libraries and cinemas – one is a public amenity, and the other is a commercial entertainment venue. In everyday cultural provision, cinemas and libraries present the most distributed and, not surprisingly, most accessible spaces in terms of attendance by a larger population group than other less-available cultural facilities (Evans 2008). Unlike cinema, where audience profiles show a rapid 'age-decay' between younger and older consumers, libraries also maintain a fairly even spread of age group usage and higher proportion of lower socio-economic group usage. Despite cumulative reductions in the funding and therefore opening hours of public libraries (threatening long-standing minimum standards of provision), in the UK, 39% of the adult (16+) population still use libraries and this rises to 50% in more deprived areas of the country: 'Libraries offer more than just books, CDs and DVDs. They have become the portal to a whole range of material for education, entertainment and self-improvement' (DCMS 2014, 10). Libraries also offer a good litmus test of participation of a local population through the library card with user address details. Whilst the full range of services actually used is not distinguishable (and is therefore understated), this does provide a clear indication of the geographic catchment in terms of households and therefore socio-demographic analysis. An example of catchment and population usage is illustrated through an analysis of library provision in Gateshead, north east England (Figure 2.4). This reveals which areas of the borough attract comparatively higher and lower library usage in proximity to library facilities. Local provision is clearly important to actual usage, but variations are apparent between areas where library facilities are similar. This information in turn can be analysed by household and demographic profiles and can suggest where there may be gaps in provision and where accessibility, literacy and quality issues may be constraining or encouraging usage. Like bookshops, particularly independent and specialist, the death of libraries has been foreseen for several years, with home delivery and digital alternatives cutting into their core service – book and

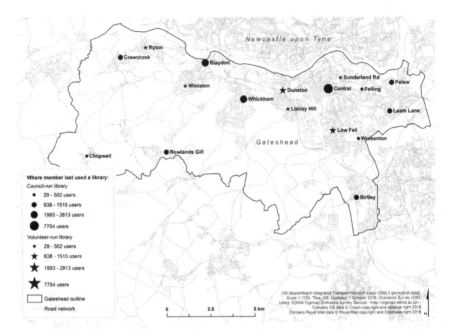

Figure 2.4 Gateshead Borough Libraries: Location, Governance and Where Members Last
        Used a Library

Source: Delrieu and Gibson (2017, Open Access).

tape/CD lending, information services and children's provision. However, embracing digital book downloads operated through library membership schemes has widened their appeal and usage again, although of course not the physical building. This has, to a certain extent, freed up space so that today, local libraries – some rebranded as *Idea Stores* or part of arts centre and civic hubs – host young people studying and using these facilities as a buffer between home and school, as well as a range of classes, spaces and exhibitions – from pre-school to pensioners. 'At every stage, the contexts – spatial, political, economic, cultural – in which libraries function have therefore shifted; so they are continuously reinventing themselves and the means by which they provide those information services' (Mattern 2014).

As well as national and museum library *grand projects* in Paris and Amsterdam, major design-led library buildings have opened in city centre sites, for example in Birmingham, Mexico City, Washington D.C. and Seattle – reaffirming their civic (and architectural) value (Worpole 2013). Their nature distinguishes them from most other cultural spaces: 'A library is a very specific sort of building. A building where you collectively do something individual. With a theatre, it's the other way round. There you look individually at something collective' (Arets 2005, in Worpole 2013, v). As architectural critic Boddy noted (2006): 'It seems evident that the building that will come to emblematise the beginning of a new century of public architecture is not the latest Kunsthalle by Hadid, Holl or Herzog & de Meuron, but rather Rem Koolhaas' Seattle Central Public Library' (in Bagwell et al. 2012, 17). The reason why libraries still have a clear civic edge over the proliferation of art galleries and museums of recent years – often in the name of urban regeneration – is because they continue to provide a much richer range of public spaces than these other forms of cultural provision, public or private (ibid, 18).

Cinema

Cinema provides another locally accessible cultural facility, albeit also declining in supply – like libraries also subject to high levels of substitution from competing forms of cultural consumption, notably TV followed by video/DVD and now digital platforms (PC, mobile/tablet streaming). Cinema, which itself had usurped the music hall and other spaces for popular entertainment (and re-occupying some of these former entertainment buildings), has been in long-term decline. In the UK, cinema attendance peaked in the 1940s (fuelled by the War years) at 1.6 million but, by 1960, had decreased to 500,000 and by the turn of the twenty-first century to only 142,000 per year. Provision is of course variable, the USA has more cinemas than Germany, the UK and France combined (Sassoon 2006), and cinema-going reflects this, with average annual cinema attendance in the USA twice that of the UK and France (four vs. two times per person). The advent of streaming films on home and mobile screens and the drastic impact of the Covid-19 pandemic on out-of-home activity from 2020 also saw a further decline in both attendance and the number of movie theatres, with the UK Cineworld chain closing over 120 of its cinemas, and in the USA, a 60% drop in cinema-goers.

Despite this decline, cinema still maintains the highest level of attendance at cultural events – 68% of the population in England, compared with 48% attending 'any performance in a theatre', and the only such activity where attendance was actually increasing (ACE 2010). Black and Asian group attendance is also higher here than in other activities – proportionately higher than White audiences (ACE 2003). Between 1995 and 2009, cinema attendance rose from 52% to 68% of the population. Like the UK government's *Taking Part* cultural activity survey (ACE 2017), this data provides no indication of the type or actual *place* of participation, whilst the frequency of attendance they count is low (it includes people who attend 'less frequently than once a year'). Cinema on the other hand has the highest frequency of reported attendance with nearly four times as many people going regularly (four times or more) than going infrequently (once or less a year). In a comparison of public engagement in culture across Europe, Brook found:

> [U]nsurprisingly, there is also a strong relationship between the number of cinemas in a country and the level of cinema attendance. The correlation between cinema attendance and the number of cinemas per thousand population suggest(s) that 83.7% of cinema attendance is explained by cinema availability, or conversely that the supply of cinemas in each country very closely matches demand.

> (2011, 22)

A novel contribution to the place–cultural experience relationship has been provided by the introduction of live theatre screenings in cinemas by the National Theatre, Metropolitan Opera (New York), Covent Garden Opera House and other larger venue organisations such as the English National Opera. For example, the National Theatre broadcasts live performances simultaneously to collaborating cinemas across the country, capturing the 'live' experience rather than just digitally recorded for later screening. One perhaps unexpected impact has been that cinema audiences are in many cases closer to the stage/performers than most theatre-goers, with cameras allowing for close-up zooming. A NESTA survey (2010) of two NT productions – Racine's *Phèdre* and Shakespeare's *All's Well That Ends Well* – at these live/screened events produced some interesting results between the cinema and NT theatre audiences. In terms of the experience itself, cinema audiences were more absorbed and had their expectations met more so than the theatre-goers, with a higher emotional response and an appreciation of a new way of seeing the work. Live cinema therefore provides both a substitute and complementary cultural experience to live theatre and provides a positive introduction to subsequent attendance at live performances (in both the cinema and theatre) with 89% of cinema audiences saying that they were more likely to attend future live broadcasts and with a large minority more likely to attend the theatre itself (34%) and other plays/venues (30%). Significantly, the live cinema screening attracted a lower-income audience and many of whom had never been to the NT. What is clear is that a familiar and accessible venue is able to attract a wider audience than the original production. In the cinema case, audiences were willing to compromise factors such as ambience and some of the excitement of the face-to-face live event for convenience, a more intimate, shared experience and closer access to the stage/action (but importantly still 'going out'). For most, local accessibility was an overriding factor in attendance (Evans 2008).

Cultural Space Perceptions

The production of cultural space is perhaps not surprisingly dominated by the political and symbolic economy (Miles 2015) surrounding the creation, *conception* and legacy of cultural facilities and their environs (the field of cultural heritage is examined at length in Chapter 4). On the other hand, the user/*usager* is subservient, or at best a subsidiary actor in an urban imaginary, suggesting that the experience of space can be unsatisfactory, inadequate or not 'representative'. As Lefebvre argued: 'this is the dominated – and hence passively experienced – space which the imagination seeks to change and appropriate', but as he also acknowledged, the specific spatial competence and performance of every member of society can only be evaluated empirically (1991, 39). Castells on the other hand seems to contradict Lefebvre on this point, observing that (Lefebvre) had scant concern for empiricism (1997 and see Elden 2004) – and there is of course little quantitative data or empirical validity underpinning Lefebvre's work in this regard (Filion 2019); however, his concepts (not 'theories to be tested') were never promoted as predictive models, or as steady states.

Observers and researchers in this field, including philosophers, sociologists and other analysts, have looked to behavioural studies, user surveys and related data collection methods to assess cultural space usage, primarily from a social profiling perspective, but empirical evidence is lacking on user and artists' perspective on these spaces and facilities. In architecture – the prime profession in space production (including landscape, urban and related engineering design) – clients are typically private patrons (including non-accountable trustees/boards of cultural organisations), city and national authorities, and this process is thus distant if not completely divorced from everyday cultural users, including those responsible for managing and maintaining these buildings and environments. Outside of perennial user surveys undertaken for funders, policymakers and to serve marketing imperatives, there is little observation or co-design in the development stage, or post-occupancy studies, despite the common experience of problematic public buildings, notably poor acoustics, lighting, safety, wayfinding/legibility, irregular layout/walls limiting picture hanging, etc., for example Guggenheim in New York and Niteroi in Rio – and a lack of fitness-for-purpose, often leading to expensive, but unsatisfactory, retro-fitting (Evans 2001, 2003a). This problem has worsened in an era of 'signature' design styles favouring irregular 'wavy' facades and footprints, materials and space layout (*vis* Gehry, Hadid, Alsop, and Neimeyer, Herzog de Meuron). Neimeyer's new-build art gallery in Niteroi for instance 'emulates a flying saucer with curved walls and no clear divide between floor and walls – a curator's nightmare. A visually spectacular landmark overlooking the bay towards Rio, it is nonetheless located in a monotonous public and inaccessible space' (Evans 2003a, 329). An observation on the professional critique of new architecture is its presentation in architecture magazines and media. Specialist architectural photographers are employed to capture images of the pristine building typically against a blue sky, *sans people*, effectively not-in-use and *a priori* its purposeful use (i.e. for what it is was both designed and funded). As Harris notes: 'the ICOM definition of a museum reveals a conception of the museum institution as an empty container with sharp boundaries between it and its users. None of its meanings seem to be drawn from an interface with people' (2015, 1). Scaling up the planning scale to urban design/master planning level, this process also combines a geometric and graphic precision through plans and models which neither reflects the real world nor the reality of the actual built environment and how it is experienced in practice over time (Evans 2015b).

Considering spatial practice, on the other hand, embodies a close association within perceived space between daily reality or routine and urban existence, that is, the routes and networks which link up the places used for work, 'private' life and leisure (Lefebvre 1991). Observing routine space usage has been a feature of urban design; for instance, the seminal work of Appleyard on *livable streets* pioneered in 1960s San Francisco (1981). Appleyard studied three similar streets with varying levels of road traffic, observing the significant variation in residents' social networks and interaction, street life and general well-being, with heavy traffic acting as a territorial divide between neighbourhoods (Evans 2014c). His methods of observation and annotative image mapping using tracing paper and capturing residents'/users' perceptions were the precursors to the *Planning for Real* and digitally enabled GIS-participation techniques, as discussed in Chapter 9. Danish architect and self-styled promoter of *Human-Centred Urban Design*, Jan Gehl, has taken this human observational approach into the public realm, now more widely applied in public space design practice. Some of his observations are apposite: 'Cultures and climates differ all over the world, but people are the same. They'll gather in public if you give them a good place to do it'; 'First life, then spaces, then buildings – the other way around never works';

> I find it striking that the quality of the urban habitat of homo sapiens is so weakly researched compared to the habitats of gorillas, elephants, and Bengal tigers and panda bears in China . . . you hardly see anything on the habitat of man in the urban environment.[11]

More quasi-scientific approaches to this subject utilise digital data analysis, notably *Space Syntax* (Hillier and Hanson 1988), describing humans as *agents* in urban space and repetitively modelling their movement habits in relation to the built environment as a predictive and urban design tool. What these observational recordings lack, however, is the cultural dimension or (remarkably) its elucidation through user involvement and expression (motivation, emotion, etc.), that is, lived experience of cultural *public* space. Most research on user interaction, for example in museums and galleries, is conducted in labs[12] (perhaps emulating evidence-based medicine and policy formulation), astonishingly without access to real artworks, without the social context of a museum and without the presence of other persons (Carbon 2017). Cultural spaces themselves seldom undergo *a priori* participatory design or post-occupancy studies (and if they did, many well-known architects would have to return their design awards) and the concept of 'design quality' is therefore also considered here, from the perspective of multiple stakeholders in cultural spaces as a first look into how user-cultural facility design interaction might be tested in practice.

Design Quality

In part to address this gap in co-creation and user-centred design, the concept of Design Quality was developed by the then Commission for Architecture & the Built Environment (CABE) and construction-industry partners in the UK. CABE was established by the Labour Government in 1999 as its advisor on architecture, urban design and public space. Its role was to influence and inspire the people making decisions about the built environment, championing well-designed buildings, spaces and places, and providing expert, practical advice to architects, planners and clients. In order to develop a more consultative and evidence-based approach to building and space design, a Design Quality Indicator (DQI) was developed which was piloted and rolled out for various public building projects. One of the first was the Tate Modern Art Museum, situated on the south bank of the River Thames.

User groups or stakeholders identified in this exercise included (multiple) visitors and regular users, project manager, facilities manager, architect, etc. – whoever are potential users of the building and space. A conceptual framework groups various design qualities across three aspects: *Functionality* (manipulation of space); *Build Quality* (manipulation of matter); and *Impact* (effect on the mind and senses), with *Functionality* assessed in terms of users, access and space; *Impact* in terms of form and materials, internal environment, character and innovation, and urban and social integration; and *Build Quality* in terms of performance, engineering systems and construction. This triangulation is important since without each element a space could be highly functional and a quality build, but to a user be soul-less, lacking in character and integration (both internal and external), a not-unfamiliar outcome in many modern cultural buildings and public spaces. Each of these factors are then translated into a Likert-style questionnaire (seven-scale rating for each question from *Strongly Agree* to *Disagree*), as suggested in Table 2.5, the answers to which can be weighted according to the relative importance of each theme.

The results can then be presented in terms of scores/ratings and graphically using Radar or Spider charts for each user/type (Figure 2.5), enabling a comparison to be shown between each of these stakeholders or users of the space. In this pilot, the results show marked different scores between visitors, architect and project manager. This analysis offers the opportunity for different stakeholders and users to

Table 2.5 Design Quality Indicator Questionnaire Extract

	Functionality
Q	**USE**
1	The building works well
2	The building easily accommodates the users' needs
3	The building contributes to the efficiency of the organisation
4	The building enhances the activity of people who use it regularly
5	The building provides good security
6	The building is adaptable to changing needs
7	The lighting is versatile for different user requirements
8	The layout allows for changes of use
9	The heating, ventilation and IT installations allow for changes of use
10	The structure allows for changes of use

	Impact
Q	**CHARACTER AND INNOVATION**
1	The building provides a sense of security
2	The building lifts the spirits
3	Visitors like coming here
4	The building reinforces the image of the occupier's organisation
5	The building is widely acclaimed for its quality
6	The building has character
7	The building makes you think
8	There is clear vision behind the building
9	The building's design and construction contributes to development of new knowledge

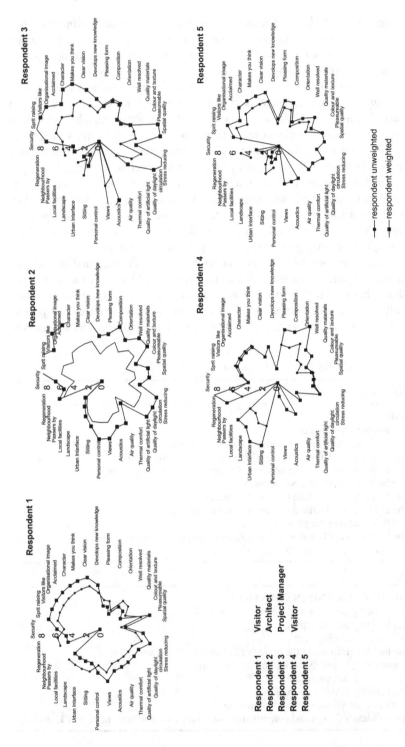

Figure 2.5 DQI Pilot Individual Impact Results

better understand the perspective of others, and if carried out iteratively at various stages of space creation (commissioning and plan/design stage, building, occupation and post-occupation) allows for adjustments to be made to the design, mix and priorities between these elements. These would be likely to change, of course, as familiarity and usage (positive and negative) are reflected in the user experience over time.

Given the hierarchy of cultural space creation (Bourdieu 1993), this approach – clearly adaptable in terms of the space design elements and judgements around their relative importance – has the potential for improving the user perspective and power over cultural space production and consumption. A particular community of interest in this field is artists, who may be the prime agent in the cultural experience whether exhibiting, performing or otherwise inter-acting with the public through their creative work, either through object/art work or in person. They are also the most absent in space production, although Lefebvre saw them as important (if largely passive) inhabitants in representational, lived space, in particular through interpretation and manipulation of systems of images, symbols and signs (Lefebvre 1991), rather than as prime users or co-designers.

Little Frank

Artists, of course, have sought to subvert the art institution through their work (within and outside of the gallery), although the pattern has seen such subversion absorbed into the institution itself, including commissioning artists to re-hang collections (Malone 2007, 16). An artist who has directly engaged with the contemporary art museum is Andrea Fraser (USA). Fraser is associated with a third generation of practice-based artists who focus on institutional critique, emerging from the 1960s and 1970s, as a reaction to the growing commodification of art and the prevailing ideals of art's autonomy and universality (Fraser 2005a). Building on earlier conceptual and site-specific art, this critique centres on the disclosure and demystification of how the artistic subject as well as the art object are staged and reified by the art institution. In a project at Guggenheim Bilbao (designed by Frank Gehry), Fraser 'performs' to the official audio guide – which opened with 'isn't this a wonderful place? It's uplifting. It's like a Gothic cathedral' – producing a seven-minute video: *Little Frank*, documenting her unauthorised intervention using hidden cameras (Fraser 2005b). During her visit, she raptly listens to the audio guide and experiences what can be described as an intense identification with the museum, as the recording rambles on about the glories of this revolutionary architecture, never mentioning the art it contains. Fraser's face expresses a range of exaggerated emotional states. When the guide discusses how the great museums of previous ages made visitors feel as if there was no escape from their endless corridors, Fraser frowns and looks pensive. When she is then told that, at the Guggenheim, there is an escape, she smiles and appears reassured but soon furrows her brow when the guide admits that 'modern art is demanding, complicated, bewildering', she quickly bursts into a grin of relief when she is told that the museum tries to make you feel at home, so you can relax and absorb what you see more easily. The less-than-subtle implication here is that instead of providing a refuge for contemplation, the museum now moves away from discussing art to turn narcissistically to itself and its affective architecture, physically and emotionally overwhelming the visitor with its spectacular spaces and grand scale (Malone 2007, 1–2).

Inter-Cultural Spaces

From a Western and generally westernised perspective, cultural spaces, whether distinguished and designated as such, or seen as distinct from everyday space and experience, can be characterised historically as White spaces, more specifically as White-European spaces. As a reflection of multiculturalism in cultural and social policy, particularly in cosmopolitan cities with diverse populations and both new and established diasporas, spaces for non-European arts and traditions have emerged at varying scales from *Grand Projets* such as the *Arab Monde* in Paris, Theatre of the New World in Montreal (sited in the allophone area of this linguistically divided city), and the National Museum of African American History and Culture, Washington D.C., to ethnic arts and cultural centres in larger cities. Governance and curation have also gradually been led from within membership of these communities (in particular, promoted by cultural agency diversity policies), with centres working with artists of African and Asian descent, for example Iniva (the Institute of International Visual Arts) in London. On the other hand, festivals such as the Zinneke Parade held in Brussels every other year since *European City of Culture* 2000 was established with the aim of connecting the many different cultures, communities and districts within the city, and to reconcile the Brussels population with its shared identity (this city in particular has suffered from 'white flight'). Mainstream cultural venues have also adopted what we can refer to as normalised multicultural programming as part of their portfolios. It is fair to say however that Black arts organisations and venues have suffered from the vagaries of funding regimes, insecurity of tenure and a lack of access to the networks and resources that many other cultural groups take for granted (Evans and Foord 2000, 2006a), as noted in the following cases.

Whilst cosmopolitan cities with diverse cultural populations, past and present, have implicitly or explicitly adopted multiculturalism, certainly in the case of public cultural space and cultural programming practice, this more passive state, at its most reductive, can reinforce the separation from the 'other', perpetuating a form of post-colonial orientalism (Said 1979). As Prigge maintains, a post-modern *Culture Society* or *Kulturgesellschaft* has the 'tendency for social space to differentiate between a plurality of lifestyles that are symbolised as "fine distinctions" in differential representations of urban lifeworlds' (2008, 50). However, Bourdieu's version of 'lifeworld' – or *habitus* – consists of the things we end up knowing without having to think how we came to know them (1984):

The lifeworld is not particularly explicit. It is a set of habits, behaviours, values and interests that go without saying in a particular context. Knowledge of the lifeworld does not have to be taught in a formal way. We learn how to be in the lifeworld just by living in it. The more other cultures and cultural forms and expressions are experienced over time, in our 'own' cultural spaces, the more these interests would be expected to embed and deepen understanding and value-systems.

(Kalantzis and Cope 2016, 115)

Multiculturalism and cosmopolitanism have also been positioned antagonistically; 'the former imagined as a project of preservation of cultural difference and the latter as transcending or eliminating such difference in the name of global citizenry' (Harris 2009, 25). But a dialectic between multi- and intercultural space is perhaps too binary, likewise between passive and active cultural engagement. So you could argue, perhaps optimistically, that a continuum from multi- to inter-cultural, and passive to active – would both lead to the other, with the opportunity and cultural space to do so (Williams 1961). Intercultural space implies this more active relationship where communities develop a deep(er) understanding and respect for all/other cultures, with a focus on the mutual exchange of ideas and cultural norms. Richard Sennett uses the example of the Greek Agora, where people can consider views other than their own as suggested in Aristotle's *Poetics* (Sennett 1998). This difference of view Sennett translates as *identity* in today's society, with its focus on race, gender and class, but

Aristotle meant something more by difference, this also included the experience of doing different things, of acting in divergent ways which do not neatly fit together, with the hope that when a person becomes accustomed to a diverse complex milieu, he or she will cease reacting violently when challenged by something strange or contrary.

(Sennett 1998, 19)

Cultural spaces that represent this intercultural spirit can therefore serve as 'safe spaces' for such interaction and mutual exchange. The location choices can however influence if not dictate altogether the degree of interaction and possibilities for cultural exchange. This is illustrated by Sennett, again, who served as a planning commissioner in New York, where he was involved in plans for creating a market to serve the Hispanic community of Spanish Harlem.

We planners chose to locate *La Marqueta* in the center of Spanish Harlem twenty blocks away, in the very center of the community. We chose wrongly. Had we located the market on that street, we might have encouraged activity which brought the rich and the poor into some daily, commercial contact. Wiser planners have since learned from our mistake, and on the West Side of Manhattan sought to locate new community resources at the edges between communities, in order, as it were, to open the gates between different racial and economic communities. Our imagination of the importance of the centre proved isolating; their understanding of the value of the edge and border has proved integrating.

(Sennett 2008)

Another example reflects the decision-making arising from the relocation of cultural spaces through a rational, but ultimately discriminatory process. In this case, two established arts organisations in Birmingham, UK – the Ikon Gallery (an artist-led gallery space established in 1965) and the Black arts centre, The Drum (opened in 1998 it established itself as the United UK's national centre for Black British and British Asian arts), were both successful in being awarded UK Lottery and European funds to relocate and upgrade their buildings (Evans and Foord 2000), but with sharply contrasting treatment and outcomes. The former was re-sited in the downtown central business and entertainment district, with surrounding café culture, and the latter in a less salubrious and non-central location in Newtown, Aston, 2.5 km north of the city centre. The justification for these different location decisions, as Symons observed: 'reveals the need for micro-environmental factors to be taken into account when planning urban investment for white and non-white audiences' (1999, 723). In this sense, the process and eventual location decision by the city strongly influence image, access/usage, markets and consequently the viability of cultural organisations and expose the differentiated values ascribed to certain cultural practices over others, even where they exist within the same art form or genre and within the same city and cultural policy regime (Evans 2001). Following financial difficulties and then liquidation, the Drum closed permanently in June 2016, a pattern seen with several Black arts organisations over the past 25 years (e.g. the ill-fated Black Arts Centre at the Roundhouse),[13] whilst the Ikon gallery continues to go from strength to strength in its new home, a gothic former school building near the regenerated city centre. Black cultural space has proved to be insecure space, despite multicultural policies and sentiments.

Intercultural City Space

The notion of the Intercultural City (ICC) is one that has been promoted by the Council of Europe with over 170 European cities adopting ICC policies and strategies, validated through compliance with a detailed questionnaire-based ICC Index underpinning these normative policies and practices, including towards public culture and space. Intercultural city spaces are first those where members of different ethnic and minority groups co-habit simultaneously and safely – the most common notion of 'shared space' – and the one which is most commonly observed and recorded in such public amenities as community gardens, parks, libraries, street markets and festivals and multicultural centres. Second, those spaces in which members of different ethnic and minority groups are encouraged directly to interact with each other through activity programmes, design or events emphasising interaction between participants. The sentiment being that, when shared space stimulates interaction between groups, this can induce a sense of belonging (Bagwell et al. 2012).

An example of a nascent intercultural space is provided in a review of ICC public spaces, including their safety and management (ibid.). In Rotterdam, *Schouwburgplein* is one of several city centre squares designated as an important focus for cultural activities for the whole city. Its stark urban design, designed to reflect the port, has been controversial. However, the square's 'cool urban' image and central location have made it a popular meeting place for young people from a variety of different backgrounds. *Schouwburgplein* or 'Theatre Square' is situated in the heart of Rotterdam, minutes from the central station, close to major shopping streets and flanked by the City Theatre, Concert Hall, Rotterdam's largest movie theatre complex, and a variety of cafes and restaurants. The square is located above an underground parking lot and is raised above street level as a result with an unusual surface made using light -durable decking and man-made materials. It consists of a central void with most activity taking place around the perimeter in the various cultural venues, cafes and restaurants. Custom-made seating is provided along one side. The square's most prominent feature is the four iconic crane-like hydraulic lights that can be interactively altered by the inhabitants of the city (Figure 2.6). These together with the hardscape surface are designed to be a reflection of the Port of Rotterdam.

Schouwburgplein was designed to be used as an interactive public space, flexible enough to accommodate a variety of different uses during the day, evening and different seasons of the year. By raising the surface of the square above the surrounding area, a 'city stage' was effectively created for festivals and installations. Regular cultural events, including music and dance, are held in the square and attract diverse audiences from across the city and beyond. During the day, the ramped roof entrance to the underground garage is used for skate boarding; other areas become an informal playground or football pitch, and the seating area provides a relatively tranquil area for shoppers and workers to take a break from work or the hustle and bustle of the surrounding shopping streets and offices. The location of the square, close to Rotterdam Central station, the shops and the cinema (rather than within particular community neighbourhoods, cf. Sennett [mentioned earlier]), means that it is an ideal spot for a rendezvous with friends and is used as such by people from all over the Netherlands and visitors. The square particularly attracts young people, including those from different ethnic backgrounds who

Figure 2.6 Schouwburgplein, Rotterdam
Source: Photograph by the author.

come to skateboard, play football, meet friends, pose, or chat up others. Being some distance from their home neighbourhoods, these young people are also away from the prying eyes of family and fellow community members (Bagwell et al. 2012).

Lamont and Aksartova (2002) call this 'ordinary cosmopolitanism', which 'has been utilised to elaborate the experiential and affectual dimensions of openness to the other' (Harris 2009, 205). Intercultural space would therefore represent a distinction from both home and 'everyday' lived space. In this respect, this would in turn influence the location, programming, content and identity of cultural space (including buildings and surrounding space and public realm/open space), an *assemblage* which is open to such interaction and engagement with the other and the new, where users can 'create a space of expression beyond institutions, at the level of the person, and can act as mediators, thus paving the way for the "shared public space(s)" necessary for intercultural dialogue' (Council of Europe 2008, 47).

*

Whilst attendance and engagement in dedicated arts and cultural spaces are generally a deliberative act, specifically situated in space and time, events and festivals have a different temporal and spatial dimension which can involve casual or even passing interaction, as well as concentrated observation and active participation in a range of outdoor places and venues that can be spread over a few hours or several days. Large-scale often-one-off events entail major construction of venues and facilities, the legacies from which add to the landscape of cultural centres discussed earlier, whilst others celebrate and elevate existing venues, spaces and routes. Community festivals, on the other hand, both cement and keep alive local and regional traditions, and new forms and themes also arise through thematic and local festivals organised to mark a community's identity and defence of cultural space that may be under threat or change. The phenomenon of events and festivals as examples of cultural spaces are therefore the subject of the next chapter.

Notes

1 https://montreal.ca/en/topics/activities-arts-centres
2 Although in the nineteenth century, opening up art galleries was also a means of social control, supposed to model respectable behaviour to urban working classes and provide them with alternative healthy leisure habits (Bennett 1988).
3 The Universal Declaration of Human Rights 'where everyone has the right to freely participate in the cultural life of the community, to enjoy the arts and share in scientific advancement and its benefits' (United Nations Human Right 1948).
4 MAC, founded in 1962 as the Midlands Arts Centre for Young People, developed the novel concept of a permanent arts centre for young people offering practical experience of a wide range of art forms. In an 8.6 acre site in Cannon Hill Park, a number of buildings, including studios and two small theatres, were completed by the mid-1960s, together with an outdoor Arena theatre. When MAC opened its creative resources included a professional theatre company as well as one of the UK's few full-time companies of puppet makers and puppeteers.

5 In 1960, playwright Arnold Wesker had a new idea for spreading the best of culture beyond the elite: 'Centre 42 will be a cultural hub, which, by its approach and work, will destroy the mystique and snobbery associated with the arts . . . where the artist is brought in closer contact with his audience, enabling the public to see that artistic activity is part of their daily lives' (Centre 42 Annual Report 1961–1962). Wesker had criticised the Labour movement for neglecting the arts. The 1960 Trades Unions Congress passed a resolution on agenda item 42 to conduct an enquiry into the arts and Wesker's initiative took the name Centre 42.

6 www.frac-centre.fr/_en/art-and-architecture-collection/rossi-aldo/teatro-del-mondo-317.html?authID=163&ensembleID=528

7 www.designboom.com/architecture/tadao-ando-commission-mpavilion-10-naomi-milgrom-foundation-australia-03-15-2023/

8 Peter Hall had been an advocate of the highly masterplanned and enterprise-zoned London Docklands development imposed by the former Conservative government in the 1980s, as well as the M25 Orbital motorway and third airport development at Stansted.

9 A recent review of *Everyday Culture* in the UK found significant "hidden" activity which was home-based, but also collective, and not categorised by simple art or cultural forms (CCV 2022). In the USA, among adults who sang, made music, danced, or acted, 63% did so in the home, and 40% did so in a place of worship (NEA 2019).

10 Although as Stopes comments (2015): 'Consuming culture on the South Bank is the right way to behave in the space, in the same way that skateboarding is not'. http://review31.co.uk/article/view/357/the-symbolic-economy

11 www.re-thinkingthefuture.com/know-your-architects/a343-10-most-memorable-quotes-by-jan-gehl-the-humanist-architect/

12 In his study of art exhibition visitors, Carbon (2017) found that the time taken in viewing artworks was much longer than was realised in lab contexts and visitors spent more time on viewing artworks when attending in groups, as well as returning to view paintings again, which further increased the time spent.

13 https://50.roundhouse.org.uk/content-items/black-arts-centre-home-art-britain-ignores

3   Events and Festivals

Events and festivals provide a particular insight to public culture in its temporal form, at both a point in time and over time as they repeat according to local tradition (e.g. religious or pagan) and periodical cycles (e.g. biennales, art festivals), or rotate across the globe in the case of competitive, international events, notably EXPOs (International Expositions), Olympics (summer and winter) and European Cities/Capitals of Culture and their emulators such as American Capitals of Culture and UK Cities of Culture (Roche 2000, 2003). In this sense, events and festival cycles represent what Lefebvre saw as the inter-section between place and time and the essential creative energy that they engender (Lefebvre 1992) – at their most vernacular and historic, festival spaces are both discursive and lived. However, the physical and heavily conceived and contrived spaces that large-scale festivals and competitions represent make up the majority of what are considered contemporary hallmark or mega-events. Early studies into the phenomenon tended to view them as simply 'special' (i.e. not regular/annual) large-scale events. However, subsequent studies (Hall 1989) also identified shorter-term staged events, such as carnivals and festivals which can be of significant economic and social importance, and which may not only serve to attract visitors but also assist in the development or maintenance of community or regional identity (Getz 2012). The term 'hallmark event' is not therefore confined to the large-scale events that generally occur within cities and major towns. Community festivals and local celebrations can also be considered as such in relation to their regional and local significance, highlighting the importance of the economic, social and spatial context within which these events take place (Evans 2020a).

The imagined and real sites they occupy – from the *a priori* location decision, event duration and post-event 'legacy' stages – raise questions regarding the nature and rationale for these events, and the extent to which their production is really cultural in intention. This is particularly relevant for events which are used for area regeneration and placemaking (Evans 2020a). Their antecedents in the eighteenth and nineteenth centuries, for instance, reveal their origins as national trade and colonial efforts, marking a divide between art and craft/trade, but whose place-based impacts have had lasting impacts on their host cities. The late nineteenth century also saw the creation of two contrasting but enduring events, the recreated modern Olympics (now accompanied by a Cultural Olympiad – Garcia 2017)

DOI: 10.4324/9781003216537-3

and the Venice Biennale (Gold and Gold 2020). One is a global sporting road-show moving to different cities every four years, and the other is a festival rooted in place, alternating each year between art and architecture events (effectively an 'annual'), but presenting a contemporary curated art and design mileux to an international audience, against the backdrop of the permanent Giardini site made up of national pavilions from the past (Evans 2018a).

This chapter will therefore consider the phenomenon of events and their production and consumption from the perspective of place and from both historical and contemporary perspectives. This includes iconic event 'cultural legacies' such as *Albertopolis* and *Olympicopolis* in London; mega-events as tools for regeneration; late night *nuite blanche* festivals that extend the museum and street life experience; competitive cultural events and their corollary where rival cities have been unsuccessful in the bidding process, but continue with alternative strategies; and finally in contrast, local community festivals that accommodate arts activism and social engagement.

The roots of many cultural festivals can, of course, be traced to rural and festival towns that host and produce traditional or 'everyday' events based on local agricultural, religious and landscapes (or a combination of all three). These often grew in size and significance as populations expanded, tourism widened its reach and urbanisation intensified, and where the rural–urban interface increased. For example, the bull-runs associated with festivals in Pamplona and other cities in Spain, France and also in Mexico and England, and London's longest running Bartholomew fair, held between 1133 and 1855, which saw livestock driven through the city to the meat market at Smithfields located just outside of the city's jurisdiction. Originally a cloth fair, it outgrew its original site and guild control to become an international cultural event until it was suppressed for its perceived debauchery and disorder by Victorian prudes. The re-creation of cultural festivals has also inevitably followed diasporic migration, with major cities not only hosting ethnic festivals across the year in settled neighbourhoods, but also outgrowing these areas, as festivals achieve large event, that is, touristic status. The question of whose cultural identities should be celebrated is therefore becoming fraught, as so many cultural groups lay claim to event space and, with the city having to limit its street festival weekends and street closures (Evans and Foord 2006a, 72), the demand for ethnic festivals has stretched city authorities such as in New York where weekend road closures have proliferated and a ban on new festivals enforced. The effect of this is felt most by new migrants who are less able to assert their claim than long-standing groups with more recognition and influence (Evans 2001).

More established examples include *Diwali Festival of Light*, *Zinneke Parade* in Brussels, and *Carnival Mas*(querade) – the latter event originally celebrated in Trinidad and then Rio, but since the 1960s with London and Toronto hosting large-scale carnivals both attracting over one million visitors each August. In the latter case of Caribana, the festival organisers were forced to relocate the event from its downtown community and symbolic finale at the City Hall to the Lake shore, with barriers separating onlookers from participants, whilst London's Notting Hill Carnival has resisted repeated attempts to be moved from its traditional west London streets and squares to a more controlled, but soul-less Hyde Park. Festival space is not there-fore necessarily benign but can be both contested and fluid. The current social pro-duction of multicultural city space, represented in both the semi-permanent-built environment and the temporary spaces of festivals, does not make cultural expres-sion easy. There are few mechanisms available to ethnic minority communities to influence the design and form of city space, and the opportunities to take control of its public spaces and streets are declining. Likewise the opportunities to shape everyday multicultural encounters, where street markets and neighbourhoods with an ethnic identity have been challenged through regulatory controls and 'improve-ments' in the urban fabric, including legislation constraining public gatherings, particularly around symbolic cultural sites. There is instead a growing trend in 'cosmopolitan' cities to create contrived ethnic quarters (Chapter 5). By enacting planning mechanisms to regulate frontages and street furniture and insisting on the use of decoration embodying stereotypical symbols of particular ethnic identities (e.g. in Little Italy, Chinatown and Banglatown), *faux* neighbourhoods are created as business and tourist destinations. These spaces are often animated by staged eth-nic festivals, funded or sponsored by local businesses, local authorities and other charitable grants and are often presented as cultural showcases for visitors (Evans and Foord 2006a; Shaw 2012).

Contemporary events are therefore a particular planning and political challenge, encompassing all scales of cultural and design practice and serving as an exemplar of a global imaginary realised in local space. They combine placemaking with national and city branding, manifested through extravagant buildings, facilities, transport infrastructure, costumes, uniforms and slogans – all of which present challenges given their high-profile, physical footprint and cost/revenue (Edizel-Tasci, Evans and Dong 2013). The production of mega-events in particular is reflected through the iconic buildings and components that make up the visual feast that these events seek to engender – conceptualised through the masterplans and computer-generated visions that are employed to project the event site into the future.

Expos

World Expos are one of the most enduring and largest events staged in modern times, taking place normally every five years and typically lasting six months. As architecture critic Dejan Sudjic observed: 'the expo is to the city what fast food is to the restaurant. It is an instant rush of sugar that delivers a massive dose of the culture of congestion and spectacle, but leaves you hungry for more' (1993, 213). In contrast to the functional and highly engineered sports stadia and infrastruc-ture required to support sporting mega-events, which apply façadist and irregu-lar roof lines to mask their box-like interiors, the international exhibitions and festivals have generated a temporary and national pavilion style which begs the question, what is their purpose? The long-established example of this is the Ven-ice Biennale Giardini. In a luxuriant 43,000 m² garden facing the Venice lagoon – commissioned by the Emperor Napoleon I and designed in the typical English garden style by the Italian landscape architect Giannantonio Selva in 1807 and completed five years later – 30 national pavilions have been built over time, with the aim of showcasing the best of each country's art and architecture during the Biennale events. The Central building, originally the Italian pavilion, was converted into a 3,500 m² venue in 2009, and accommodates one of the two curators' exhibitions, the other being located at the Arsenale, a disused naval base, whilst each national pavilion features its own art or architecture exhibition. Since 1980 Venice has held the Architecture Biennale alternating with the Art event, joining the already established Contemporary Music, Film, Theatre, and most recently, Dance festival. The pavilions are a novel architecture exhibition in themselves (Figure 3.1), with constructions built after designs by celebrated architects, such as Josef Hoffmann (pavilion of Austria, 1934), Gerrit Rietveld (Dutch pavilion, 1953), Carlo Scarpa (sculpture garden of the Central pavilion, 1952, and pavilion of Venezuela, 1954), Alvar Aalto (Finland pavilion, 1956), and Sverre Fehn (Nordic countries pavilion, 1962), among others. The latest pavilion built at the Giardini is that of Australia completed in 2015 and converted into a swimming pool for the 2016 Biennale.

International EXPOs, on the other hand, provide the opportunity for temporary national pavilions and installations which attempt to capture the essence of a country's culture and both cultural traditions and contemporary goods, but which

Figure 3.1 Hungary Pavilion, Giardini, Venice

Source: Photograph by the author.

can degenerate into miniature tourist board theme parks (Ley and Olds 1988). The exhibitions themselves adopt high-minded themes and sub-themes, in recent years with a strong environmental sentiment. This is ironic given their land-hungry, costly and unsustainable nature (Evans 2018). For example, Aichi in Japan themed its 2005 EXPO around an eco-city concept and a 'rediscovery of Nature's Wisdom'. Japanese attendance at these EXPO events indicates their popularity, which is not matched outside of Asia – 64 million visits were made to the 1970 Osaka EXPO and 20 million to Tsukuba in 1985, with over 70 million visits made to the Shanghai EXPO in 2010 which adopted the strapline of 'Better City, Better Life'. The 2015 EXPO in Milan also took an explicit ecosystems theme: 'Feeding the Planet, Energy for Life', with nine themed zones: Bio-Mediterranean, Arid Zones, including food chains: Fruit and Legumes, Spices, Cereals and Tubers, Coffee, Cacao, Rice:

> [T]he idea of EXPO Milano 2015 is to create an Exposition in which every project, every piece of content, every part of the program has been developed with the goal of making visitor experience the central focus. The approach also makes the themes clearly perceptible.
>
> (Vercelloni 2014, 5)

This thematic design allowed smaller countries to be clustered by theme, rather than be marginalised and overshadowed by the larger country pavilions. Milan's regional growth aspirations, reflected in the EXPO's expansive incursion out of the city itself, also remind us that 'mega-events cannot be considered anymore as "exceptions", but as recurrent episodes of wider urban change processes' (Di Vita 2020, 80).

Like the majority of hallmark events such as the *European Capitals of Culture*, host cities and nations foot the bill for the honour of holding these costly extravaganzas, so variations are seen in the relative wealth and money spent in each case. Over 240 countries participated in the 2010 Shanghai EXPO but only 145 five years later in Milan – geopolitics is therefore a factor with presence in China more important (trade and cultural relations) than that in Italy. An indication that the EXPO phenomenon may be peaking is illustrated back in Osaka where fifty years on it is preparing to host the event again in 2025, however only 130 country pavilions are so far committed and with spiralling costs forecast to exceed £1 billion the government has not ruled out cancelling the project. Countries therefore choose to be included in EXPO events, although the absence of a national pavilion signifies a lack of recognition and participation within the international milieu – better to be seen in a minimal or low-cost installation than not at all. In a departure from the norm, the UK opted to commission Thomas Heatherwick for the 2010 Shanghai EXPO to design not a building, but a dandelion-shaped 'seed cathedral' covered in 60,000 crystalline spines which were tipped with tiny lights to illuminate the structure. The sculpture won the Bureau International des Exposition's (BIE) gold award for best pavilion design. Each spine contained a different seed from Kew Garden's Millennium Seed collection in London, an initiative that aimed to collect and conserve 25% of the world's seeds by 2020. The seed cathedral was dismantled and the rods were donated to various charities, schools and the World EXPO Museum, which opened in 2017, another legacy from the 2010 Shanghai EXPO. Generally, however, EXPO site design tries to respond, often too literally, to these aspirational themes, whilst national pavilions seek to promote their caricatured cultural identities within these high-minded thematic priorities (Figure 3.2).

Figure 3.2 Dutch Pavilion, Shanghai EXPO 2010 and Italian Pavilion, Milan EXPO 2015

Source: Photographs by the author.

Event Master Planning – Past and Present

A feature of contemporary mega-events is their growing scale. As noted already, established festivals have spread their footprint and reach in their respective cities, but it is the expansive regeneration plans and aspirations that now drive host cities and regions to use the once-in-a-lifetime opportunity to create new urban villages, districts and extensions to the city. The mega-event thus provides a political and financial incentive to accelerate urban development as part of grand placemaking schemes to achieve growth for rising populations and for new education, cultural and play zones for the post-industrial city (Evans 2015b). This is seen in the case of Barcelona following the 1992 Olympics and as host to a self-styled mini-EXPO Forum in 2000 which showcased the regeneration of the former industrial area of Poblenou into a high-technology/R&D zone. This houses a relocated six sepa-rate college buildings of Pobra Fabra University of Art & Design, connecting the new high-rise extension area of commercial offices, retail malls and apartments, in the last piece of the post-Olympics jigsaw. London's Olympic Park likewise now hosts the *Olympicopolis* development first envisioned by then London Mayor (and subsequently dishonoured Prime Minister) Boris Johnson, and modelled sentimen-tally on the *Albertopolis* development in South Kensington, the legacy from the 1851 Great Exhibition, and now the museum and education quarter of west Lon-don (Evans 2020c; Gold and Gold 2016). *Olympicopolis* has since been re-named less glamorously *East Bank* by the Labour mayor Khan, referencing the South Bank cultural quarter on the Thames, and which comprises satellites of University College London, as well as the relocated London College of Fashion, the V&A East Museum & Archives, BBC Music Studios and Sadler's Wells new East Dance Theatre – all moving to their new neighbouring waterside buildings between 2023 and 2025 (Evans 2020c).

Great Exhibitions and Culture-Led Regeneration

The touchstone 1851 'Great Exhibition of the Works of Industry of All Nations' had generated a surplus of over £186,000 (worth £25 million today). This was primar-ily used by Prince Albert to purchase nearly 90 acres of land stretching southwards from Kensington Gore in the area to be known as *Albertopolis* to house the com-plex of museums, colleges and institutions (today this includes the V&A Museum, Science Museum, National History Museum, Royal College of Art, Imperial Col-lege University, Royal College of Music, the Albert Hall and the Royal Geographic Society) that, it was hoped, would celebrate the interplay of the arts and science and technology and their application to industry, which the Exhibition itself had sought to catalyse and represent. In 1862, its sequel, the International Exhibition, was built on the newly constructed Cromwell Road, although less popular and less financially successful than its predecessor. Both exhibitions attracted over six million visits equivalent then to one-third of the country's population, triggering the building boom which transformed the area from a rural landscape of farms, nurseries and market gardens into a prosperous metropolitan area. This compares with the 6.5 million visitors to the Millennium Dome Experience which ran for 12 months in 2000 (Evans 1995), and the ten million visits to the Festival of Britain which ran from May to September 1951. Unlike in 1851 with the Crystal Palace in Hyde Park rebuilt and extended in its new south east London location, the 1862 Exhibition structure was demolished two years later, to eventually be the site for the Natural History Museum which opened in 1881, further expanding London's prime museum quarter.

A Victorian example then, of what today is termed event or culture-led regeneration (Evans 2005, 2015), is being played out in the Olympic Park and adjoining neighbourhoods of Stratford and Hackney Wick in east London. Like the Queen Elizabeth Olympic Park, this Victorian placemaking *grand projet* was enabled by new transport links with the inexorable expansion of the railways from the 1830s, which led to Lord Kensington selling his defunct canal to the railway entrepreneurs connecting London to Birmingham and beyond and to south of the river. In the early twenty-first century, over £6 billion in public money had been invested in extending the channel tunnel rail link, the East London line and Docklands Light Railway station upgrades to Stratford with over £3.5 billion already having been spent in extending the Jubilee Underground Line (JLE) from central to south and east London. This public largesse effectively saved the ailing Canary Wharf commercial property development, connected the Millennium Festival Dome at Greenwich peninsula, and fuelled a two-thirds increase in the number of visitor attractions and accommodation provision in the new JLE corridor between 1993 and 2000 (Evans and Shaw 2001).

A fundamental difference between then and today, therefore, is that the contemporary urban cultural renaissance (Evans 2001) has been predominately publicly funded by the state (i.e. taxpayers), albeit for substantial private benefit to commercial developers, landholders and investors (Evans 2020c). The 1851 Great Exhibition and subsequent exhibitions had in contrast been financed by private subscription, whilst entrepreneurs and industrialists had developed much of the infrastructure, and until this first 'industrial turn', supported much cultural provision for workers and urban settlements where they operated. So the claim that the success of Albertopolis and the legacy of cultural institutions illustrate the strength of Victorian policies on cultural democracy and the importance of state support for the arts (Roth 2015) is debatable. Indeed, whilst Prince Albert through his presidency of the Royal Society of Arts promoted the Great Exhibition, Queen Victoria on her accession had stopped the annual grant to the patent theatres, an early example of a cut in state arts' funding (Pick 1997).

Around this time, Thomas Carlyle among others had also begun to use the word 'industry' in a new, pejorative way – no longer referring to 'hard work and diligence', but to more organised forms of mechanical reproduction transforming our way of life, captured in Carlyle's term 'the industrial revolution', signifying not just the changing nature of work and economy, but social and cultural relationships and values.[1] This contrasted for example with Wedgwood's employees who worked a shift, not piece-work system, with both men and women trained in a range of recognised art and crafts skills (his pottery bore the distinctive stamp of the individual designers who worked with him), but with few of the production line efficiencies that were to be adopted in the mechanised industrial revolution that the Great Exhibition had both celebrated and promoted (Pick and Anderton 2013).

This was reflected in the categorisation of the 13,000 exhibits on display into raw materials, machinery, manufactures and 'fine art', which arguably marked a further shift in the sub-division of art, with fine art equated to the high skills applied to industrial products, as Pick maintains: 'the Great Exhibition hastened the creation of the two distinct categories of fine art – one "design", forward looking, and the other, "crafts" concerned with the skills of the past' (Pick and Anderton, 125) – former director Roy Strong had proposed to suffix the V&A, the 'National Museum of Decorative Art & Design'. As the V&A's first director, Henry Cole, concluded after the 1851 Exhibition: 'the history of the world records no event comparable in its promotion of human industry' (ibid.). However, one of the key visions for the event was that of 'unity' (another key term used by the Exhibition's proponents was 'civilisation', directly equated with industrial development): 'as if the Fine Arts, the new factory systems, science and commerce were still united industriously pursuing common goals' (ibid.). This had been a founding aim of Prince Albert and Henry Cole on the formation of the original 'Museum of Manufactures', and the application of fine art to manufacturing, which was created to house objects purchased from the 1851 Exhibition, and its later manifestation as the Museum of Ornamental Art (then the South Kensington Museum), to today's V&A Museum – in this transition, effectively blurring the distinction between art and science/technology.

Ideologically, the Great Exhibitions acted as a break point, in both the narrative of museum history and in the wider economies of material culture, presenting 'an opportunity for the new "Products of Industry" to merge with art and entertainment in a previously unimaginable new leisure space' (Cummings and Lewandowska 2000, 54). The Exhibition event itself had exercised price and social discrimination – first open to the leisured classes, the 'great folks' who paid five shillings for admission in relative comfort, or on other days to the working classes who came in much greater numbers (as many as 70,000 daily), but paying only one shilling, with the 'working bees of the world's hives being offered inferior facilities to view an exhibition supposedly glorifying their industriousness' (ibid.). The high arts-popular dialectic was also evident from observers such as Ruskin who 'dismissed the Crystal Palace as being neither a palace nor of crystal, but merely a glass envelope to exhibit the paltry arts of our fashionable luxury' (Evans 2020c, 38). Cultural production had been led by an expanding consumption market which drove the crafts and design manufacturing industries, as Store Smith observed: 'the middle class family now possesses carpets and hangings . . . and not a few of our London middle-class tradesmen possess a better stock of family plate and linen than many a country squire' (1972, 20).

The event had also generated what today would be a case of arts, or rather event sponsorship – the drinks company Schweppes had paid £5,500 for the Great Exhibition franchise and sold over one million bottles of their soft drinks – today Coca Cola, one of the main franchisees of the modern Olympics, and Schweppes brand-owner, contributes £64 million every four years to the event. During the 17 days of the London 2012 Games, Coca-Cola served 18 million of its sugary products. Another Olympics parallel is the impact from this hallmark event on established arts providers and merchants who suffered from their trade being diverted to Hyde Park for the 1851 season, as the Exhibition soaked up both public support and spare capital – a complaint also aimed at the London 2012 Olympics where the Government had diverted significant Arts and Heritage Lottery funding from designated 'Good Causes' and other parts of the country, to the capital-hungry Olympic enterprise in east London (Evans 2020c).

The imperial Great Exhibitions also further cemented the dual cultural hegemony that had been reinforced as the nineteenth century progressed with English (and French) cultural production starting to dominate publishing, theatre and crafts, as early cultural globalisation was fuelled by the expanding Empires and industrialisation (Sassoon 2006). The World's Fairs had originated in the French tradition of national exhibitions, a tradition that culminated with the French imperial Industrial Exposition of 1844 held in Paris, following on from a series of great exhibitions which had begun in the seventeenth century with exhibitions of works of art (and thus predating the first Venice Biennale in 1895). These were held biannually from 1667 to 1793, and from 1793 to 1802, they were annual, before reverting to a biannual pattern under the French Empire. Along with the art exhibitions were exhibitions of French manufactured goods, which although not international in scope were the more direct ancestors of the universal exhibitions (Daniels 2013). As Greenhalgh observed: 'the importance of these for Government at the time was evident; they were no mere trade fairs or festival celebrations, they were outward manifestations of a nation attempting to flex economic, national, military, and cultural muscles' (1988, 6).

The Great Exhibition in London had also been preceded by two smaller exhibitions which were staged by the Royal Society of Arts in 1844 and 1849: 'wedding high art with mechanical skill' (Pick and Anderton 2013, 135). The 1851 Great Exhibition was also an explicit advert for the British export industry, with the majority of exhibits coming from the Empire and predominantly Britain's manufacturing cities; for example, Sheffield had 300 exhibits, from railway springs, vices, anvils to newly designed fenders and kettles. As Herbert Read observed in *Art and Industry*:

> Those splendid institutions in Trafalgar Square and South Kensington, now treasure-houses which attract pilgrims of beauty from every corner in the world, were first conceived as aids to the manufacturer in his struggle with foreign competitors. The National Gallery and the Victoria & Albert Museum in London, prototypes of similar institutions all over the world, were not founded as Temples of Beauty but as cheap and accessible schools of design.
>
> (1932, 138)

The surplus generated from the event it was hoped would help to set up museums in the leading industrial cities, but the balance was only sufficient to purchase the South Kensington Estate. The resentment and lack of support for the London 2012 Olympic event from the regions that was evident from the national population attitude surveys conducted by government (notably in Scotland – Evans 2016a) was also present in 1851, with London capturing the legacy and financial benefits. Provincial industrialists responded to the London event with estates purchased and used for 'pleasure parks', such as Adderley Park in Birmingham (1856), the Arboretum in Nottingham (1852) built in the style of Crystal Palace, the Manchester Exhibition at Old Trafford Park (1857) and Kelvingrove, which housed the 1888 Glasgow International Exhibition, and which in turn raised funds for the Kelvingrove art gallery and museum.

Spatial Narratives

EXPO sites, with a curious legacy of abandoned sites and a few permanent pavilions, can also take decades before they are fully redeveloped, such as in Lisbon (1998) and the UK Garden Festival sites at Gateshead and Liverpool, whilst others struggle to reinvent themselves, notably Seville (1992) and Hanover (2000). This German EXPO presented a confused theme that resulted in a little over a half of the forecast 40 million visitors and a deficit of over $600 million. The site continues in its original form as an exhibition site, although a new centre of information technology, design, media and arts has been located there. Several national pavilions were retained but are in a state of disrepair.

In planning terms, the emerging practice of master planning event sites now leads the spatial design process, within which architecture, landscape and other design activity are situated and subservient. So that whilst, in the past, architecture would have been the prime design profession providing the design concepts, iconic images and themes, it is urban design and master-planning firms that visualise the mega-event and major regeneration schemes worldwide. The practice of master planning as a new hybrid spatial and communication design process locates these special events as urban imaginaries (Çinar and Bender 2007) through a convergence of visual and virtual culture. Gonzales refers to the tension between the need for a 'spatial fix', on the one hand, and the reality that scales are socially constructed and therefore not fixed but 'perpetually redefined, contested and restructured' (2006, 836), on the other hand. Master planning therefore seeks to capture spatial design and land use configurations at larger scale than traditional architectural design or even town planning. This hybrid practice – attempting to integrate architecture with planning through urban design – thus follows a hierarchical design iteration: master plan-urban design-quarterisation-zoning and, only then, individual sites, buildings and structures which populate the futuristic graphics and fly-throughs used to envision and promote mega-event sites (Evans 2015b). Cuthbert perhaps naively reminds us that 'urban design is not merely the art of designing cities, but the knowledge of how cities grow and change . . . we must go beyond abstract social science into the realm of human experience and the creative process' (Cuthbert 2006, 1). In reality, however, this creative 'virtual' visualisation practice also underpins the process of placemaking that now drives the urban design and branding imperatives that foreshadow mega-events. These large-scale projects can therefore be seen not as the 'stormtroopers' but as the 'aerial bombardment' of gentrification, accelerating new private housing and city extensions and the displacement of residual industry and incumbent communities, including low-cost artist studios (NFA 2008). In practice, the human dimension in this process is limited to CGI-generated bodies and avatars.

Contemporary mega-events are thus creating a new landscape in their respective cities. The practice and primacy of the architect have been overtaken in this field, with the urban designer and master planner creating the canvas within which building, landscape, interior and product designers compete for attention. Visual meta-themes and styles are set at this level, which actually limits creative scope and individuality, but which nonetheless requires a complex response to these overarching imperatives. Festival sites have provided often singular legacies in the past (e.g. Eiffel Tower), but the contemporary mega-event is both more expansive and expensive – and, as a result, controversial and contested (Cohen 2013; Powell and Marrero-Guillamon 2012). This is evident in cities that have actively chosen *not* to bid for these extravaganzas, such as Toronto and Rome, echoing cities that have resisted the Guggenheim franchise (Evans 2003a). In Hamburg for example, the citizenry rejected through a local vote, the idea of hosting the 2024 Summer Olympics at HafenCity, opting instead (although not via democratic vote) for a similarly expensive concert hall: the Elbphilharmonie or 'Elphi' (Kuhlmann 2020). Designed by stadia architects Herzog & De Meuron (v *is* Beijing Olympics 'Bird's Nest' and Allianz Area, Munich), this controversial project did however emulate its Olympian counterparts by costing more than four times its original budget of €200 million.

Notwithstanding this occasional resistance to the mega-event, cities in developing regions, notably the Middle East, such as Dubai and Qatar (e.g. FIFA World Cup), vie for international cultural and related trade events and satellites of national museums, biennales and institutions. The opportunities and challenges are high, not least because of the huge budgets involved and the global reach and coverage they can generate but also because the legacies they produce – physical, recorded and in collective memory – can be significant and symbolic. Mega-events can therefore be seen as grand placemaking schemes for the twenty-first century, drawing on their boosterist past and further extending the hard branding of the city: 'creating a form of Karaoke architecture where it is not important how well you can sing, but that you do it with verve and gusto' (Evans 2003a, 417).

Hold Back the Night – Late Night/Nuit Blanche Festivals

In contrast to construction-led mega-events, cities large and small explicitly use their critical mass of existing cultural assets (Keegan and Kleiman 2005) as part of competitive city and placemaking strategies, and increasingly in order to celebrate their cosmopolitanism whilst at the same time maintaining their particular heritage legacy, image and brand (Evans 2022). A particular manifestation of this festival city phenomenon (Palmer and Richards 2010; Gold and Gold 2020) is special events which seek to reconcile this traditional heritage (e.g. museums, historic sites, intangible heritage) with new programmes that appeal to a wider range of visitors and participants, with the hope that they both extend their cultural (tourism) offer and market, spread the geographic and temporal distribution of cultural activity, and revitalise their cultural and historic venues. These events also present an opportunity for cities to reaffirm their identities and status to residents and visitors alike.

A novel example of this phenomenon is *Nuit Blanche* or Light Night events and festivals which have multiplied in European and North American cities and further afield over the past two decades, from Tel Aviv to Tokyo, and from Miami to Madrid (Evans 2012a). Their origins vary, but light-night/all-night events have been associated with religious and cultural festivals (e.g. Diwali Festival of Lights, Bhuddist Lantern Festival), predominantly held in the early autumn, with early examples of light-night events identified with St Petersburg's White Nights cultural festival held over several weeks, and Berlin's *Lange Nacht* or Long Night of Museums, with museums staying open until 2 am twice a year. Since Paris inaugurated the *Nuit Blanche* in 2002, a movement of such festivals has developed and gained momentum. In Paris, Christophe Girard, Deputy Mayor, first proposed this event one year after taking office, and the concept was rapidly adopted in other cities with five capitals – Paris, Rome, Riga, Brussels and Madrid – organising an exchange programme for artists, each welcoming an artist/theatre company from each city. The following year, Bucharest joined these six cities to formulate a shared artistic project based on the creation of a 'lounge area' in the heart of each Nuit Blanche (Jiwa et al. 2009). In many cases, there has been a rapid build-up of activity, scope and attendances at these events in only a few years, as they move from local, special interest from artists and area/business improvement initiatives, then adopting a wider festival district and route, and finally to international appeal and status.

This evolution of *Nuit Blanche* events has therefore spawned a European network with common aims, and this reflects their cooperative nature and a celebration of the expansive European Project. For instance, the 'twinning' of Paris and Rome; European Museum Nights and networks of national events across several cities, for example, in Ireland: Dublin – Cork, Limerick, Waterford, Galway; France: Paris – Amiens, Brison, Metz; Italy: Rome – Specchia, Genoa; and the UK Light Nights cities of Belfast, Birmingham, Bournemouth, Brighton, Leeds, Liverpool, Nottingham, Sheffield, Stoke, and Kirkaldy and Perth in Scotland. A European Charter was created to promote and coordinate the *Nuit Blanche* branded event, which subsequently spread to Canada (Montreal, Toronto, Halifax), the USA (Atlanta, Chicago, Santa Monica/Los Angeles) as well as Peru (Lima), Israel (Tel Aviv) and Malta (Valetta), with Nuit Blanche events now held in over 120 cities. The majority of these events reference Paris; use the *Nuit Blanche* tag; and in several cases French cultural organisations are active in funding, sponsorship and event promotion. New York inaugurated its first Nuit Blanche bringtolightnyc arts festival in October 2010, followed by Melbourne in 2013, with the city's galleries, theatres, music venues, shops and major cultural institutions showcasing the city's art and culture, food and wine, fashion and sport. This first Melbourne White Night was attended by an estimated 300,000 people, the largest event of its kind in Australia. Continuing this trend, *Noche en Blanco* was held in Santiago, Chile, and in Bogota, Colombia, during 2015. From surveys conducted during these events, satisfaction with their quality was high, for example in Rome 90% (42% 'Excellent', 48% 'Good') and Dublin 65% 'Very Satisfied'. Nearly half of visitors participated in two or more activities and nearly 80% travelled by foot or public transport. Visitors to these late-night events were primarily local/city residents, with a growing out of town and international tourist or visitor to the larger city festivals (Evans 2012a).

Late night cultural festivals have also evolved independent of the *Nuit Blanche* brand and network, notably in Atlanta, Chicago, Dublin and Copenhagen. Here, local cultural development agencies rather than city mayors organise the event, albeit with city council funding and infrastructure support. In the UK, which has not created a *Nuit Blanche* branded event (adopting the Parisian version would not have been politically palatable), a Light Nights network nonetheless effectively emulates *Nuit Blanche*, although lacking the all-night programme of events and extended opening. The reluctance of these regional cities to go the full 'all-nighter' reflects disquiet with the perceived negative social effects noted earlier. The UK Light Nights events thus emphasise a feeling of community sprit; an awareness that the city is united and that there is a sentiment that everyone has something in common – bringing social causes to people's attention; allowing people to experience different forms of art and entertainment and to experience things that they would not normally; and creating an awareness of what the city has to offer to residents and visitors alike. Following light-night events in Birmingham, Leeds and Nottingham, several more cities now offer their residents and visitors the opportunity to experience their city at night from a fresh perspective. The sentiment of reclaiming the night in these city centres is also strong: 'Light Night gives the general population the chance to "take back" the city from the demographic group that normally occupies the city in the evening and at night' (ATCM 2009, 1).

Features of these *white night* events are illuminations on buildings and light installations, including fireworks displays, late night opening of museums and galleries and in some cases performing arts venues, parks and gardens, sports facilities, as well as live events in major squares, stadia and waterfront sites. Public transport – normally free/low cost – is extended into the early hours along festival routes and to venues, with additional bus and tube/tram services to cope with the extra demand. This festival event is also increasingly a vehicle for new artists and award schemes, including biennales, and in some cases, children and young people's events and participation, as well as community and local-area development. The events are normally free, although in some cases a combined ticket is purchased for a series of museums or galleries and transport (e.g. €1 ticket in Berlin). All of these events, however, seek to exploit and rediscover the evening and late night economy and respond to concerns over safety and crowding out of visitor activity by mono-use of city centre spaces associated with extended alcoholic drinking in bars/clubs and associated anti-social behaviour, in an attempt to 'reclaim the night'. For example, Berlin and Copenhagen have longer established late night cultural events and festivals that have successfully offered a different experience, audience and geographic spread to that offered through late night bar/club activity. Berlin in particular has benefited from a liberal licensing regime and responsive approach to music and other creative uses of space, many created in interstitial zones and spaces in former East Berlin (Evans and Witting 2006). Dublin, suffering capacity constraints of a small over-heated late-night Temple Bar zone (Montgomery 1995), subsequently also widened its coverage, adopting an area, or 'quarter' approach, which is a familiar strategy evident from other Nuit Blanche festival cities (Evans 2012a, 2020b).

In May 2009, museums in the UK joined with French museums in hold a late-night-opening weekend. Paris started its *Nuit des Musees* in 2007 with Museums of Modern Art, Bourdelle, Victor Hugo's house and several others opening from 6 pm to midnight, with talks, installations such as deckchairs in galleries, writing and drawing workshops. Since 2006, the Secretary General of the Council of Europe has been promoted to the Museums Open Night which takes place concurrently in the signatory countries of the European XII Cultural Convention, including the UK. The initiative promotes Europe/European cultures and provides the opportunity to attract a wider public, particularly young people. Over the first weekend, Museums at Night was held in the UK with museums and heritage attractions opening until midnight: 'For us it was a really successful way of promoting the [National] Gallery to new audiences and those who cannot normally visit during the day' (Culture24 2009), where a talk by the Velvet Underground's John Cale and a late opening of the popular Picasso exhibition attracted more than 300: 'the noticeable arrival of a more fresh-faced bunch than museums normally welcome was a particular trend. Generally we noticed a much younger crowd were in throughout the night' (ibid.). In some cases, museums stayed open throughout the night, for example, Tate Modern from 5 pm to 5 pm the next day, whilst more than 120 people visited Florence Nightingale House in London after dark.

Open Studios

At a yet-smaller scale, production districts, as we have seen in the case of other cultural spaces, represent symbolic, historic and important elements in a cultural ecosystem and economy (Chapter 5). Often hidden physically, operating in re-used buildings and former manufacturing spaces, as well as 'at home' (Holliss 2015), their locations have limited scope for showcasing outside of factory outlets and off-beat boutique shops and temporary pop-ups. In some respects, their 'rights to the city' are neither facilitated through formal cultural institutions or spaces nor acknowledged in mainstream economic and market systems. Local cultural trade shows often start off very small and are self-organised, before becoming absorbed into wider and larger-scale events. In Berlin for example, the annual *Design Mai* Festival ran for three years, organised by a local society coordinated by seven voluntary members. The initiative started out as a magazine – 130 open studios participated over a two-week period in May each year. Some locations provided a venue for several design presentations, whilst a Showroom offered a retail opportunity to purchase direct from designers/creators. The central festival venue was the Forum in Berlin-Mitte, with an auditorium for workshops, lectures and presentations. *Design Mai* transformed into an international as well as a Berlin-wide event – in 2005, over 12,000 tickets were sold and the *Design Mai* website received six million hits, with the initiative now absorbed into the Berlin Design Week which still runs during May each year.

Open studios, where artists and designer-makers open their otherwise closed workspaces to the public, also took off in London, spreading to other regions and then internationally. Whilst major art and design fairs (i.e. sales-based showcase exhibitions) are held annually, these are highly selective and competitive, and expensive for small designer-makers. Often held in 'posher' parts of the city (e.g. *Chelsea*, London and New York), they are also not part of an authentic local cultural production space or community. Partly in response to this hierarchy, *Hidden Art of Hackney* began in 1994 as a small event to promote quality work by over 40 east-end artists/designers from their studio base (Foord 1999), designed and promoted as a studio tour and map, a simple format which is widely used today in festival and open studios marketing. A decade on, it developed into a unique network which has promoted and supported over 1,800 designer-makers and built links between the creative and manufacturing industries. Open Studios events have now spread to several London boroughs and other cities (e.g. Toronto – Gertler 2006), with design exhibitions, once 'trade-only' now transformed into cultural events – *Designer's Block, 100% Design*, and contemporary art shows (e.g. the annual *Frieze Art* fair which receives 60,000 visitors), and fringe 'off-piste' events with art and design students end-of-year shows, music and club gigs and publishers extending the scope and duration of these ostensible product shows (Evans 2014c). Not surprisingly, these are now international events that spawn other creative production showcasing for example the London Design Festival receives over 400,000 visitors, a quarter from overseas. Their uniqueness lies in consumers coming to the place of production, that is, studios and workshops, which is increasingly rare, not least in the placeless digital media sphere and exclusive worlds of fashion and music. Open studios are now held in counties and towns throughout the UK each year.

Architectural design too, which in terms of engagement has remained restricted by museumification, drawings and models, has followed the open studio format with festivals combining open architecture studios with installations, lectures and children's events. In its first year (2004), London's Architecture Biennale attracted over 25,000 people over the opening weekend to what would have normally been a deserted area of the city (Figure 3.3). In the 2006 biennale, a route across London responded to the theme of *Change* with *starchitect* Norman Foster driving sheep across the Millennium Bridge, which was co-designed by his firm. *Open House* similarly taking place over a weekend each year gives access to working heritage and other buildings of architectural interest. In these ways, the everyday culture and landscape of the city are celebrated and accessed, in contrast to the increasingly deleterious experience at staged tourist and heritage sites.[2] Today the annual London Festival of Architecture ('pre-Covid') receives over 800,000 visitors during its month-long programme. As we have seen also in the case of Fashion and

Figure 3.3 London Architecture Biennale, Clerkenwell, London
Source: Photograph by the author.

even Digital industry trade shows (Chapters 6 and 7), there is a latent demand for closeness to creative and cultural production and producers that cannot seemingly be met by other forms of communication or experience, a corollary perhaps to the thirst for live music events in an age of streaming.

Thematic and large-scale festivals tend to occupy particular and controlled parts of the city, whilst arts festivals provide an opportunity for existing venues and companies to raise their profile and celebrate place/places – from the large-scale Edinburgh Festival and prestigious arts festivals that occupy cultural tourist itineraries, to perennial international film, theatre, dance and art events. A far cry from these 'party-fests for the well-heeled' and cultural tourists, local events bringing together local artists, galleries and communities many of whom do not attend formal cultural events or venues – and/or lack access – take place in neighbourhoods, in local parks and around community centres and hubs. Local arts festivals, which are either based in one venue with a varied programme or across larger and smaller venues, can include dedicated arts centres, theatres, community venues, parks/ open spaces as well as pubs, libraries and even shops. Examples include Literary and Poetry festivals, Art 'Opens' (which include home-based artist spaces), Music (e.g. jazz festivals) and Crafts and Food festivals. The festival format is also used in more activist ways, combining the local arts festival with issues focused on resistance to – for example development, gentrification and socio-political themes such as climate change and racism. Indeed early 'free' festivals from the late-1960s provided a 'safe' opportunity for alternative lifestyles, so-called counter-culture and expression to be promoted and experienced. Two seminal free music events took place in 1969: Woodstock, in Bethel upstate New York, a three-day *Aquarian Festival of Peace and Music*, and *The Stones in the Park* (Rolling Stones headlining in Hyde Park London). Both attracted nearly 500,000 audiences and marked key moments in music history. In 2017, the Woodstock festival site became listed on the National Register of Historic Places. At the other end of the spectrum, a month on from Hyde Park, David Bowie organised and starred in the Beckenham Free Festival in his local park, as an extension from his weekly Arts Lab (see Chapter 2), which took place each Sunday evening in a nearby pub, The Three Tuns (now demolished to make way for a chain pizza restaurant).

The appeal for mass gathering at music-oriented events has not diminished; however, these now resemble controlled and commercially driven jamborees – from Glastonbury, England (weekend ticket £340), Coachella, California ($550) to Primavera, Barcelona (€325). In the UK alone, eight new-purpose-built arenas are in various stages of completion, not just in cities without a large-scale venue, but second and third arenas in London and Manchester (Aviva Studios) with capacity for over 20,000. On the other hand, a new arena in Bristol will house three venues, rehearsal and TV production rooms, education programmes and community stages for local bands, whilst at the smaller end of the scale, smaller venues are declining in number (including pubs, traditionally home to folk, rock and jazz clubs). This also reflects the shifting economic landscape of music, due to declining income, with record/CD sales usurped by online streaming and single-track downloads, leaving live gigs as the main income-generator for bands. Rising fees and event costs – and therefore ticket prices – have seen rock and pop emulate the lyric arts that, since the 1990s, have seen venue prices for opera, theatre and dance rise in real terms. To use economist's jargon, exploiting audience consumer surplus and 'willingness-to-pay', although in the Arts case, not actually increasing their audience size, but succeeding in getting their increasingly exclusive audience to pay more for the privilege (Evans 1999b).

European City of Culture

Designated and self-styled cities of culture represent the dialectical universal and particular – demonstrating on the one hand their European-ness and reflection of the European (Cultural) Project (Evans and Foord 2000), and on the other, their unique, indigenous or at least endogenous cultural assets and identity. In Lefebvre's terms, thus simultaneously reflecting perceived and conceived space – a reminder that these conceptual states can exist both at the same time through varying perceptions, vantage points and representations, and which can be mutual or conflictual. The *European City of Culture* competition was conceived at a time when the city was again perceived (politically identified as the 'urban new left', Bianchini and Parkinson 1993) as a place of culture, style and artistic excellence, and when industrial production had declined both economically and symbolically. The scheme was proposed by the Greek culture minister in 1984, and the following year, Athens was selected, followed by Florence in 1986. Fourteen European cities hosted annual cities of culture festivals, which were used to shoe-horn new and upgraded cultural facilities (e.g. the makeover of the Prado Museum, Madrid in 1992) and re-package their cultural itineraries. In the millennium year with growing pressure from cities seeking culture city status, nine cities were selected which included, for the first time, the east European cities of Prague, Krakow as well as non-EU member Reykjavik (as Istanbul was selected in 2010). This signalled not only the extension of the European project east and north but also the aspiration of former soviet bloc countries to participate in the network of culture cities and, by definition, claim/regain their place in the European Renaissance. Over 60 cities/city-regions (e.g. Ruhr) have been awarded ECoC status, and despite its lack of novelty and European provenance, the city of culture concept has also been adopted across the Atlantic where a competition for *Cultural City of the Americas* saw Merida, capital of the Yucatan province in Mexico, host the first festival in 2000. Helsinki also shared Capital of Culture designation in this year. The city had embraced the cultural industries and arts flagship strategies, which included developing a cultural production zone centred on the Cable Factory and music conservatoire, and a cultural consumption quarter around the Glass Media Palace. This 'Palace' – a row of shopfronts and cafes backed by a central coach station – was an inheritance from the ill-fated 1930s Olympic Games, converted to an art house cinema, art book shops, cafes and media production facilities to form part of a cultural triangle with the new museum of contemporary art and multiplex cinema nearby. The obligatory contemporary art museum, in this case *Kiasma* ('a crossing or exchange'), designed by American architect Steven Holl and like Guggenheim-Bilbao, hosting a private, imported collection, somewhat curiously 'seeks to redefine the art museum as an institution, shifting from the image of elitist treasure house to public meeting place' (Verwijnen and Lehtovuori 1999, 219). In other ECoC cases, the impact on cultural provision and activity has been marginal, low cost and largely celebratory rather than regenerative in intent, for example Cork, Ireland, and Leewaurden, the Netherlands. Recent studies also suggest that whilst cultural festivals have many merits and may contribute to well-being, they tend to have limited, temporary impacts on local economies and creative industries' growth, despite the explicit economic development goals of both the bidding cities and in the assessment of the awarding agencies (Nermod, Lee and O'Brien 2021).

The bidding process for the renamed European *Capital* City of Culture parallels the build-up and optimism surrounding EXPO and major event competitions driven by geopolitics, commercial sponsorship and the high risk associated with aspiring culture cities which lack the scale, image and infrastructure necessary to achieve success. This winners and multiple losers game has created a growing tier of peripheral and regional cities that repeatedly enter such competitions and justify major public investment in new venues and transport in terms of the regenerative benefits of branding that will hopefully accrue to them. Cities such as Hamburg

(Hafencity), Manchester (East) and Toronto (waterfront) have repeatedly staked the regeneration of major sites on such failed bids (e.g. Olympics), often despite popular resistance, whilst others have been burdened by the financing and failure associated with undeveloped post-event sites, in some cases long after they occurred, for example, Montreal's 1967 EXPO and 1976 Olympics – an ongoing extra tax on cigarettes; and Sheffield 1991 Student Games – the closure of local and community sports and leisure facilities, and higher entry fees for new facilities (Evans 1998b). As in the *Grands Projets* of Paris and elsewhere, such as Guggenheim Bilbao, these cultural flagships have in practice crowded out cultural provision in other parts of the city and region and substituted a particular version of culture – the heritage and monumental art museum and performing arts and sports complex – for contemporary cultural activities and facilities which reflect more diverse, contemporary and participatory arts and entertainment activity (Evans 2003a).

On 10 January 2009, Liverpool held the official closing event under its European Capital of Culture (ECoC) programme exactly a year on from its inauguration. This event also saw the transition from Liverpool's Year of Culture '08 to Year of Environment '09 (repeated again in 2019), and a simultaneous event in the Austrian city of Linz to which the Capital of Culture mantle passed, along with Vilnius, Lithuania. An estimated 60,000 people congregated at the Pier Head as well as at the Albert Dock and Wirral bank, for a celebration that included sing-a-long, firework displays, street artists on illuminated bikes and light projections onto a famous refurbished new museum building making up this World Heritage city. Cities of Culture (CoCs) associated with city branding, re-imaging and regeneration using culture and flagships to soften and celebrate urban renewal efforts can also be characterised as the festivalisation of city development (Palmer and Richards 2010) and of identities – political, economic and community – and as one element in a continuum of thematic 'Years' and 'Cities of . . .' – in which Culture is just one manifestation. However, findings from the Liverpool'08 Impact study revealed the mixed and marginal effects felt by residents, with 'ambivalence about the likelihood of sustained benefits resulting from ECoC' and 'concerns over the sustainability of retail and leisure developments and viability of new city centre apartments' (Impacts08 2009, 13). Critically 'most people did not feel that ECoC would benefit either them as individuals or their neighbourhoods' (ibid, 12). Here as elsewhere (Evans 2011), a case of undervaluing community culture and amenities, and the exaggeration of event activities and interventions, are felt by local people.

The European City/Capital of Culture project has an extended but crowded history, as they have 'doubled-up' and multiplied over 40 years. Evidence of their contribution to culture and placeshaping is not surprisingly mixed, given the range and size of cities participating and their respective starting points and levels of cultural investment. In some respects, they have become ubiquitous and no longer 'special' events. To date, no cities have hosted more than once, although Brussels (co-ECoC in 2000) have bid again for the 2030 event. Research into ECoCs is also extensive, over and above the generally celebratory (and mandatory) post-event reports issued by host cities and sponsors (including the EU) and a small library of good practice guidance (see Palmer 2004; Cox and Garcia 2013; Richards and Marques 2015; Nermod, Lee and O'Brien 2021), but these are nonetheless dominated by 'impact studies' – economic, social but less so cultural – and characterised by a dichotomous culture of optimism by city proponents (city hall/mayors, cultural agencies, EU) and, on the other hand, more pessimistic analyses from academic and local community perspectives. Cities that have bid or been shortlisted but failed in the final award have also continued their cultural development strategies either with their festival budget promises intact (e.g. Maastricht 2018 – Evans 2013b) or, in many cases, claiming that the bid process itself established improved networking, sense of pride and identity, even if this could not be maintained post the bid award stage (Richards and Marques 2015). Similar claims for the soft benefits of bidding are also made by Leeds, who were the victim of 'cancel culture' – as the UK left the European Union, the city's bid for the 2023 ECoC was withdrawn. Despite this, the city continued with *Leeds 2023*, neither European nor UK City of Culture, but maintaining a five-year cultural investment programme and event year: 'to reshape the image of Leeds as a city with something important to say in terms of culture'.[3] What shape that might be or take is however, as yet unclear.

UK Cities of Culture

In part to encourage smaller towns and cities, particularly outside London, to host cultural festivals, and also due to the infrequency of national allocations for the European CoC, the UK Culture Ministry launched a national City of Culture award programme in 2009. With the UK leaving the European Union (and European Economic Area) in January 2019 ('Brexit'), the UK ceased to be eligible for the European CoC award from 2019, so this initiative effectively replaced the ECoC for the UK. Like their European antecedents, UK Cities of Culture (CoC) are prime examples of drawing on their tangible and intangible heritage both in order to bid for the CoC award and to programme, and to a lesser extent, develop a post-event legacy. A key inspiration for the UK programme was Liverpool's 2008 European Capital of Culture, and the UK forerunner, Glasgow in 1990. Pre-dating these cultural events, both cities also hosted Garden Festivals in 1984 and 1988, respectively (as did Ebbw Vale, Wales and Stoke and Gateshead, England), reflecting the long-term aspirations and timescale of the regeneration process which events can both punctuate and help catalyse. An indication of Glasgow's legacy is the recent award as UK's top cultural and creative city by the European Commission's Cultural and Creative Cities Monitor 2019 report which ranks 29 different aspects of a city's cultural health, including its cultural vibrancy, creative economy and ability to attract creative talent and stimulate cultural engagement, also reflected in its status as a UNESCO City of Music. Another example is Liverpool's selection to host the Eurovision Song contest on behalf of war-torn Ukraine in 2023.

The afterglow generated by these placemaking events – despite instances of negative media, local resistance or indifference and often contrary evidence to the good news advocacy – continues to attract newcomer and established towns and cities, cities that seek to repeat or recapture past event effects, or those seeking to reposition themselves or update their heritage or dated images. When the UK Cultural Ministry (DCMS) opened a call for UK CoCs, 29 candidates including several urban and rural sub-regions put themselves forward – Birmingham, Derry (Northern Ireland), Norwich and Sheffield shortlisted, with Derry eventually winning.

Derry

Derry/Londonderry was therefore the inaugural UKCoC, held in 2013, and therefore the event where longest post-event period has elapsed. As well an association with the 'Troubles' and a longer history of colonialisation and division, economic and social decline were features that the CoC also sought to address. This included the extent to which culture could be 'life and placeshaping', offering an opportunity to use the year to enable multiple readings of identity, history and place which would include a range of communities in defining and debating the traditions, histories and heritage that mattered most to them. Surveys of residents, stakeholders and visitors were undertaken with evidence of genuine transformative change regarding image improvement and civic pride, enhanced community relationships and sense of unity, shared and depoliticised spaces and cross-community attendance. Improvements in employment, social deprivation and tourism (e.g. visitors and hotel occupancy declined in 2014 and 2015) were not evident however, with a lack of demonstrable impact and legacy (Boland, Murtagh and Shirlow 2019).

As the authors conclude, aside from unrealistic expectations and targets – social, economic, consensus, histories – a more dialogic process shaped by community narratives, 'claims' and hard evidence would better offer the potential to challenge and shift the debate beyond the sterile rivalries of heritage and its misuse in the service of urban culture (Murtagh Boland and Shirlow 2017, 519). This approach has also been adopted in heritage and placemaking research across Europe in an attempt to break down the distinction between the observer and the observed, aiming for co-production of knowledge in order to deal with the complexity of urban environments and heritage (Oevermann et al. 2022). In practice, however, artistic directors and other key producers are typically hired from the outside, often after the festival has been conceived and the narrative written. This also reflects the reality, as with the majority of larger urban festivals, that these are not just imaginaries produced by a cultural and city elite, but do not engage – through co-creation or co-design of themes, programmes or production – with the supposed beneficiaries of the event and the *a priori* rationale for community benefit, beyond token citizen consultation exercises.

Hull

The first English City of Culture under the UK's programme was the port city of Kingston upon Hull which was awarded the second UK CoC held in 2017 with themes celebrating the contribution of the city to art and industry: *Made in Hull*, manifested in 11 commissions of sound and light projected throughout the city centre, attracting more than 342,000 visits over seven days in January including firework displays over the Humber Estuary and light projections onto landmark heritage buildings, and *The Blade* installation, a 75-m wind turbine, made in Hull at the Siemens Gamesa factory in Alexandra Dock, installed in the city's central Queen Victoria Square by artist Nayan Kulkarni. Intangible heritage also featured, exploring the emancipation movement and contribution of William Wilberforce, who was born in Hull, and building on the pre-existing, annual Freedom Festival that was first held in 2007. These one-off festival events of necessity draw on existing cultural assets and spaces and the extent to which they enhance this provision is an important measure of their effectiveness, for example by collaborating and attracting new users and audiences, by raising their profile and by creating new experiences in neighbourhoods outside of the designated and city centre venues and sites. These branded events do however follow a pattern, of high aspirations and excitement at award and build-up stages (including partnership building, if not genuine co-design), followed by frenetic event programming, delivery and media generation, and then post-hoc rationalisation, including commissioned impact studies generally led by local universities and consultants, based on audience and visitor surveys and feedback, and economic impact modelling (the latter notoriously exaggerating economic impact values – TBR and Cities Institute 2011). Lasting effects, tangible and intangible, are seldom captured due to the timescale and attention decay: 'Many of the most important outcomes of Hull *UK City of Culture* 2017 will only be fully assessable three, five or even ten years after the end of 2017' (University of Hull 2019, 9); and how and what might be measured and attributed are unanswered here.

Coventry

Coventry took over the mantle in 2020/2021 and staged a higher profile event, extended to 2022 as a result of the Covid-19 pandemic which limited openings and attendance at events. Coventry likewise used its manufacturing and making traditions, pre- industrial (e.g. bicycles and motor vehicles), as well as post-industrial popular music, *Two-Tone*, and it's literary past, to project its identity and programme. Event sites included Coventry Cathedral, a symbol of war (ruins and adjoining new building), peace and reconciliation, the nearby Herbert Gallery which hosted the Tate Gallery's Turner prize, and other themed exhibitions. Local arts projects in selected neighbourhoods adopted an approach of putting 'co-creation' at the centre of its programme, valuing local stories and the latent creativity across the city. This way of working sought to deliver a range of more intimate events and higher impact activities in order to generate long-lasting social value. Embedding engagement at a hyper-local level, the approach supposedly saw local communities including faith groups, community centres, libraries, schools, community radio stations, the police and local arts organisations helping to shape and design the creative programme (attributing the actual programme with this process is not however apparent). The Coventry Heritage Trust that has been gradually taking over and repositioning (renovating, re-use) key heritage buildings and spaces

from the City Council had little involvement in the Culture Festival or its programming, although as the Trust generously admitted: 'it has been greatly assisted by
the tail wind created by Coventry's year as City of Culture (Figure 3.4), which
has allowed it to achieve in four years, what it had originally planned to do in ten'
(in Evans 2022, 16). Despite an exhaustive series of impact reports led by local
universities and consultants,[4] lessons do not seem to have been learned from previous city festivals. Less than six months on from the festival finale, the City of
Culture Trust went into liquidation with debts of over £4 million (most will not be
paid, including to several arts organisations in the city), making all remaining staff
redundant and being forced to sell off its assets, despite a £1 million loan from the
City Council (£12 million was spent on events but only £487,000 earned in ticket
sales). Recriminations have replaced the euphoria and congratulatory messaging
from the city, as one of the disenfranchised local artists observed: 'it leaves behind
forced redundancies, disgruntled but mobilised creatives, a disappointed population, unpaid debts and lots of unanswered questions, not quite the three year legacy
it had supposedly planned and we had hoped for' (Manning 2023, 2).

Whether explicit or not, the decision to host a major or city-wide event is an act
of place branding (Evans 2022). The image, programme mix, theme and heritage,
tangible and intangible, are the prime elements that a city assembles in order to
plan, pitch and promote an event or festival to a range of stakeholders – internal

Figure 3.4 UK City of Culture, Coventry 2022
Source: Photograph by the author.

(residents, businesses and politicians) and external (funders, awarding bodies, press and media, and wider public). City promotional campaigns also incorporate events in order to animate what can be corporate and consumption-oriented strategies targeted at inward investment and tourists. Smith (2016) uses the term *festivalisation* as the process by which urban space is produced via the staging of events, and the development of festivals at a local level has also been adopted by business districts in order to stimulate a visitor and creative economy in off-beat or previously mono-use areas of the city, such as Chicago's *Looptopia* (Evans 2012a, 2014a). This can also draw on lesser-known heritage of an area, such as Little Germany in Bradford, UK – an area of rich architectural heritage dating from nineteenth-century German–Jewish textile merchants and celebrated by an annual festival and mural on the nearby Bradford Playhouse.

These city festivals thus take everyday, familiar urban spaces and attempt to elevate them for a short period (and perhaps paternalistically, also take plebeian communities and seek to engage them though exposure to selected cultural activities). Tensions however perennially emerge between externalised professional artistic direction, programming and the city's visitor and marketing goals, and the participation of artists, communities and cultural spaces whose own aspirations and perceptions of place maybe at odds (or not in the same 'space') with the festival roadshow. With hindsight – and a not inconsiderable evidence and knowledge base accumulated over the past 30 years (Richards, Brito and Wilks 2013; Evans 2020a) – the competitive event and festival process seem less likely to produce cultural spaces that are sustainable and valued, whilst festival hierarchies built around intermediaries increasingly resemble the *artist-as-scientist* that Lefebvre included in his characterisation of the representers of conceived urban place production, along with other social engineers (including academics and consultants), planners and architects.

*

Whilst the physical legacies of past events and the place associations with annual festivals help to define key parts of the city (including the after-use of event facilities), ongoing participation in, and perception of, events and festivals is less visible, not just in memory and memorabilia, but in the preparation, skills and cultural development throughout the year, for example in Carnival dance, music and costume workshops and making, as well as community rehearsals, programming and pre- and post-event activity. Events therefore present us with a particular physical and intangible heritage space, no less so than the historic monuments and sites that feed the *heritage industry*. The concept, practice and perceptions of cultural heritage are therefore explored in the next chapter, where: 'in the face of apparent decline and disintegration, it is not surprising that the past seems a better place. Yet it is irrecoverable, for we are condemned to live perpetually in the present. What matters is not the past, but our relationship with it' (Hewison 1987, 43).

Notes

1 www.jstor.org/stable/40002946
2 Open House carried out research among 3,000 members of the visiting public in 2008 and found that 70% were surprised by the architecture they saw; 66% said that Open House London made them think differently about London; and 24% found out more about sustainable/green design through the event. Visiting buildings was cited as the most informative and enjoyable way of finding out more about architecture (Evans 2020b).
3 https://eurocities.eu/latest/a-rising-tide-lifts-all-boats-leeds-2023-explained/
4 https://warwick.ac.uk/about/cityofculture/our-research/ahrc-uk-cities-of-culture-project/futuretrendsseries/

4 Cultural Heritage

The modern concept and practice of cultural heritage are perhaps the most histori-cally and socially produced and place-bound cultural form, directly influencing its consumption and relationship to community identity and sense of place. The insti-tutionalisation of cultural heritage can be linked with processes of urbanisation and industrialisation from the late nineteenth century, accelerated by this process and from the destruction from World War II in the twentieth century, with a response in the form of building protection. Listing and preservation of buildings and historic sites have been practiced since the late nineteenth century (e.g. Historic National Sites, Canada) and from the mid-twentieth century; national and city conservation legislation was created specifically to protect buildings and heritage quarters from modern development, for example in London (*Civic Amenities Act* 1967), Paris (*Plan de Sauvegarde* 1970) and Montreal (*Heritage Montreal* 1975), whilst the UNESCO Word Heritage Convention in 1972 kick-started the World Heritage Site listing regime. The 1970s also saw the rediscovery of the historic city as places to live and work, with communities seeking greater involvement in decision-making over development and planning, and in placemaking efforts (Hosagrahar 2017; Madgin 2021).

Heritage in this sense can be distinguished from the more anthropological con-cepts of culture and tradition – where tradition refers to the process of handing down from generation to generation some *thing, custom or thought process* that is passed on over time. Marx for example distinguished between the concept of heritage which he saw as encompassing all historic and style periods, all social formations without exception – from tradition, which is only a component of the former, defining tradition as spiritual relations and as spiritual connections between objects and circumstances (Marx 1852) – the 'wealth of ideas consolidated in the public mind, which requires a choice, acceptance and interpretation of the heritage from the point of view of certain classes, social layers and groups' (Andra 1987, 156). The built and site-based version of heritage dominates the institutional world and the formalised administrative procedures based on expert knowledge. Critical heritage studies on the other hand challenge this established knowledge on monu-ments and heritage conservation, including what even constitutes heritage, who defined it, and how and by whom historic site, memory and memorisation are con-stituted, performed and re-framed (Smith 2006; Waterton and Watson 2015). This

DOI: 10.4324/9781003216537-4

is evident today in view of the controversies over monuments and statues and the reinterpretation of art and museum collections and heritage sites with problematic provenance or 'difficult heritage' (Tunbridge and Ashworth 1996).

The beginning of the twenty-first century has therefore seen growing interest in how heritage is considered – its contribution to our understanding of the past, the way it is approached by academic research, and its place in the evolution both of landscapes and of societies. Much effort is required to give this past, and especially its material remains, the social recognition it deserves and how it can make contributions to contemporary society in terms of sustainable development, urban regeneration, architectural invention and local economies, because it means something to us, it often reflects our generation/memories and we want to keep that meaning for future generations – but what precisely does heritage mean to today's societies? What importance do we attach to the past? What influence can heritage have on the way we experience the everyday that also reflects our collective identity? These are questions raised in Bazelman's *Redefining Heritage* concept (2014) which seeks to reflect the widening of values attached to the material past. In his and others view, these values all need equal consideration in the interpretation and design of shared heritage spaces, buildings, landscapes and natural heritage (Evans 2022, 2016b).

Heritage and Place

Cultural heritage is also important in the process of placeshaping, particularly in the context of regeneration and culture-led regeneration more specifically (Evans 2005). Indeed, without the tangible and intangible legacies from the past, building or rediscovering an authentic place identity is very difficult (Evans 2022). Heritage valuation – and its measurement – also reflects the inherent selectivity often involved in heritage designation and preservation and therefore its social production. This is important in what is included (and excluded) in cultural maps and official audits of heritage and their perceived values (Evans 2008; DCMS 2010):

Heritage is by definition an inheritance from the past destined for the future. It is a product of the present purposefully developed in response to current demands for it, and shaped by that market. It follows that the past is a quarry of possibilities from which selection occurs not only or even principally by chance survival but by deliberate choice.

(Ashworth 1994, 1)

Selectivity is therefore a key to heritage planning. A dichotomy exists between the original positivist 'preservation' and the normative 'heritage' which implies a process of selection and conservation of history, memory and relics, and their interpretation for contemporary consumption. As Ashworth again maintains: 'You cannot sell your heritage to tourists: you can only sell their heritage back to them in your locality. The unfamiliar is sellable only through the familiar' (1994, 2). Heritage selection and valuation are also applied in order to prevent further decay and degradation, but the notion of the ruin should not be limited to collapsed architectural structures, instead any structure that contains gaps and absences punctuated by time also functions as a ruin (Davidson 2022, 5). This 'ruin effect' (Simmel 1958, 379) situates these remnants as essentially 'a temporal figure, the dialectic of nature and culture prompting a sense of time passing' (Davidson 2022, 5). Heritage as represented in art and architecture is also subject to assessment and valuation by the scholarly canons of art history and through codification and curation and the symbolic importance attached by heritage experts. Whilst such designation has been dominated by classical and iconic styles represented by historic monuments, castles, churches, cathedrals, palaces, museum quarters and their collections – *Grand Projets* of the past – more recent heritage has however begun to appear in designation by preservation movements.

The distinction between tangible and intangible heritage assets also emphasises the physical (buildings/sites, artefacts) and human (cultural diversity, living heritage) dimensions which are now recognised in various designations and awards by national and international heritage and human rights organisations, for example World Heritage Sites and Cities (Evans 2010), listed buildings, Conservation areas (ICOMOS, UNESCO, Historic England) and also networks such as Historic Cities, Towns & Villages and Cultural Routes. Engagement and participation with heritage in official discourses are however limited to the built/historic environment, with heritage policy and research dominated by conservation and material science (e.g. EU Framework and H2020 programmes), a field now intensified by Climate Change imperatives and heritage degradation (DeSilvey et al. 2022; Harrison and DeSilvey 2022; EC 2009).

However, international and national heritage organisations have more recently adopted what can be characterised as a people-centred approach, which also values intangible heritage. The UNESCO Creative Cities Network which awards this status to prospective cities (perversely, the UNESCO designation only allows one of these creative categories per city), covers seven cultural fields – *Crafts and Folk Art*, *Design*, *Film*, *Gastronomy*, *Literature*, *Media Arts*, and *Music* – recognising that heritage is constituted of meanings, shaped by people's perceptions, whether as objects, knowledge or practices, with 'enactment' an essential aspect of intangible heritage in the sense that this heritage exists and is sustained through the 'acts of people' (UNESCO 2001). The concept of living heritage in particular emphasises the creative and dynamic nature of heritage as continually adaptive and changing, highlighting ongoing practices as primary mechanisms for preservation and cultural renewal (Simpson 2018, 3). In Japan and South Korea, this concept has even been personified through designating 'living cultural treasures', recognising those human treasures whose knowledge and skills enable the performance and transmission of intangible heritage.

The human rights dimension has also been fuelled by indigenous community claims and campaigns to secure access and ownership of traditional lands and artefacts, including those held in western museums. This has included recognition of the integral and inseparable relationship between cultural and natural heritage, that is, cultural landscapes, and with increased diversification of values, the essential subjective nature of heritage – a 'cultural construct reflecting the diverse and localised meanings and values attached to places and objects' (ibid., 2). The first such cultural landscape to be inscribed by UNESCO in 1992 was Tongariro National Park in New Zealand, in recognition of the cultural and religious significance of the mountains for the Maori people. The attention to intangible heritage, often place-based (currently or in the past/memory), has therefore demanded changes to disciplines and institutions that had operated within the western paradigm and which equated heritage with materiality and nostalgia for the past. It is evident that 'the symbolic marking of places, the preservation of symbols of recognition, the expression of collective memory in actual practices of communication' (Castells 1991, 351) are very important in order to recognise and, if necessary, protect the identities of places.

The tangible, of course, can also be a source or an inspiration for the intangible and vice versa, and intangible legacies are often subsequently celebrated in museums, exhibitions and festivals of music, fashion, food and culture/history – for example, fashion, LBGTQ and slavery museums. However, heritage assets are not limited to the formally designated, such as listed buildings and museums, since local historic assets and practices – both tangible and intangible – are often valued higher than official heritage assets and narratives: 'people are increasingly interested in different aspects of our history that our listing and designation policies have not traditionally recognised' (Historic Environment Scotland, in Madgin 2021). This has belatedly been acknowledged by heritage agencies, such as Historic England's Heritage Action Zone (HAZ) and High Street HAZ scheme working in partnership with local communities and partners in everyday settings, and a recently launched *Everyday Heritage* fund, which focuses on working-class history and intangible heritage (Evans 2022). Local amenities such as libraries and parks also represent embedded and frequently used heritage assets and help to define a place and particular community, likewise the memories and experience derived from community festivals, key historic events, characters and industrial heritage. This more inclusive perspective on cultural heritage is therefore important, since it can better define a place and identities and capture a richer and more representative image of a past and contemporary culture of a place. In Australia for example, cultural planning guidance has encouraged co-production of heritage narratives, allowing local communities to write their own cultural histories and profiles, linked to facility maps and images, whilst a GIS-based cultural atlas in Western Sydney has created a web resource allowing the user to zoom in to images, video, audio, stories and links to documents and producing trails and tours (Evans 2008).

Whose Heritage Is It Anyway?

Participation 'in heritage' is notoriously difficult to measure. On the one hand, this is due to the loose and expansive nature of what constitutes heritage – buildings, sites, monuments, districts/quarters, towns/cities, routes – and on the other hand, the fact that most heritage assets are free to access or at least view, without controlled entry. Heritage facilities which do have controlled entry and ticketing are obviously able to measure visits/visitors (but not *non*-users), both subsidised and commercial, and libraries are a good example of collecting data on users/usage type which can be mapped by household/library member and compared across other library provision (Brook 2011), as discussed in Chapter 2. Surveys, whether site or population based, provide further analysis of users of heritage, in terms of their profile, behaviour and relationships or 'emotional attachments' (Madgin 2021) with historic places, but these are generally small and infrequent samples. Another challenge is the varying scales at which heritage is considered, notably historic environments, towns and landscapes. One consequence of this situation is that claims for engagement with heritage rank highly in official data, with 73% of those responding to the *Taking Part* cultural activity survey (DCMS 2020) say-ing that they *visited a heritage site* in the last year (compared with 52% visiting a museum). These sites included visiting a city or town with historic character (the most frequented heritage sites), a monument (e.g. castle, ruin) or a historic park or garden open to the public. The most common reasons given for these visits was: *to spend time with friends and family* (46%), followed by: *having a general interest in heritage or history* and *just being in the area* both at only 26%, with *lack of time* (37%) and *interest* (36%) the two prime reasons for not participating.

Critically, these surveys are also not place-specific – a self-defined 'visit' could have been undertaken abroad, on holiday, on a day trip to another area/city or locally. As a review of government cultural activity studies concluded (Ebrey 2016, 160): 'in measurements of "value" inherent in government research hitherto, such as those in the *Taking Part* survey and the CASE (evidence) programme – neither study gives the public a free or genuine voice' (Walmsley 2012, 330). The rigour of evidence in this field therefore tends to be lacking: 'the key ideas in the [heritage] policy literature tend to be asserted, rather than represented as derived from evi-dence, and the specific arguments generated through academic research are not vis-ible in (such) policy' (CURDS 2009). Whilst much heritage is place-based and/or associated with place, it can also be mobile and cross boundaries, such as historic routes and trails, for example the Council of Europe's Cultural Routes programme (Khovanova-Rubicondo 2012), which links cities and towns with intangible and tangible legacies such as industrial heritage (e.g. Hansa league) and intercultural city spaces (see Chapter 2).

Evidence produced through official UK data on participation and engagement in heritage also reveals key inequalities between different groups. For example, only 41% of Black respondents visited heritage sites versus 75% of white and 60% of Asian respondents,[1] whilst only 51% of those from the most deprived areas did so, compared with 83% in the least deprived areas. A similar gap is seen in heritage visits between those in higher managerial and professional occupations (84%) than those working in routine, manual occupations (62%). In the case of museums, this disparity is more marked – with only 10%–20% of the latter socio-economic group (C2DE) attending museums, whether free or charged-for, although interestingly this attendance rate was much higher (40%–50%) for new satellites of existing museums (e.g. Tate Liverpool and St Ives). This suggests that both proximity and symbolic place values attached to these museum outposts combined to attract these traditionally lower-participating groups. Those living in rural areas (83%) were also more likely to have visited heritage sites than those from urban areas (70%), an indication perhaps of a greater preponderance of traditional heritage sites in countryside locations (e.g. stately homes, castles, national parks). This is significant for cities which represent urban areas of higher deprivation and with a higher proportion of ethnic minorities for instance. This includes those with industrial legacies with important representation in popular culture – from literature, film, music to fashion and food – and diverse intangible heritage drawn from this history and from a multicultural past and present.

Engagement with heritage is, like other provision, supply-led; however, there are variations in both 'supply and demand', even in a European context. Museums, perhaps the most commonplace institutional heritage space, are a key indicator in this regard. For example, Austria has over 240 museums per million population compared with only 43 and 45 in the UK and Italy; 60 in France and Germany; and 18 in Spain. Visitor numbers at these museums reflect this level of provision with 2.8 million visits per million population in Austria; c. 1 million in UK, France, Germany and Italy; and 0.7 million in Spain (Creigh-Tyte and Selwood 1998, 152). Whilst access (including opening hours and entry fees) and proximity are obviously important, there are likely to be qualitative (collections, display, permanent vs. temporary exhibitions) and socio-cultural factors that influence museum-going habits, including education, social networks and competing or substitute cultural activities.

Bourdieu's comparative surveys of art museums (Bourdieu and Darbel 1991)[2] carried out in the mid-1960s provided an insight to the socio-economic/occupation profile and prior experience of visitors, with museums selected according to their size (number of visitors and works on display), a Delphic assessment of their quality (curating and art works), and listing in guides, with correlation (i.e. a similar hierarchy) found between these three ratings. The survey findings reaffirmed the correlation between (higher) levels of education and employment ('class') and, in this case, the frequency of museum visitation compared with those with lesser education and occupations, but, through observation, time spent and explanatory guides and tools used by different groups, their recall and naming of specific artists, and the age at which they first experienced the art museum (with family, school or other visit). This laid the foundation for the concept of cultural capital and in the context of a cross-national survey, that of national cultural capital, distinguishing between for example the Polish – who exhibited a less differentiated visitor profile across the social classes compared with the elitist west Europeans, which the authors put down to a tendency to 'compensate for the relative paucity of its cultural capital by a sort of cultural goodwill right across the social spectrum' (Bourdieu and Darbel 1969, 36). On the other hand, from the exceptionalist French perspective:

the whole cultural tradition of countries with an ancient tradition (*Ancien Regime*) is expressed in a traditional relationship to culture which, with the complicity of the cultural institutions responsible for the organization of the cult of culture, can only be constituted in its own modality where the principle of cultural devotion has been inculcated from earliest infancy by the encouragement and sanctions of family tradition.

(ibid.)

In hindsight, this was an example of cultural relativism, with its conclusions drawn from a particularly formal cultural experience, the traditional art museum, but arts attendance in Poland was in fact higher than France or the Netherlands (Sassoon 2006), but obviously within a different, but not necessarily less cultural social and educational structure. The Poles surveyed also reported the highest number of prior museum visits than French, Dutch or Greek counterparts (although like the amount of time spent, the authors put this down to over-reporting). Family, social and cultural structures therefore vary, and privileging one system over another has provided a weak foundation for cultural theory over time, as has the valorisation and selection of cultural heritage, which has been the subject of attention.

Legacies from the past, of course, continually accumulate, both marginally and sometimes radically as a result of major redevelopment, regeneration and new/extended facilities, for example, museums and galleries. Heritage protection is now afforded modern buildings and sites as the more recent past becomes 'historic', with preservation and documentation bodies such as the Victorian Society and Twentieth Century Society campaigning for protection and listing. This begs the question – how many examples of a particular architectural type or heritage building or site should we conserve? Nationally, this can be viewed rationally (and via a national planning system), but, locally, an exemplar heritage asset, if not unique, may be valued highly by residents but less so by national and professional bodies. There is in particular scope for further exploration of the links between the historic environment, sense of place and social capital, including active placeshaping projects and ethnographic studies, to better understand how the historic environment and heritage activities might figure within people's daily lives in different areas and regions of the country (CURDS 2009).

Heritage City

The notion and adoption of the 'Heritage City' tag conflates these two distinct coinages. City status involves not just size and population but also symbolic importance, with the latter a function of history and institutional and political processes. Royal charters, cathedral cities, Cities of Culture, and provincial and capital cities are all examples. Heritage, on the other hand, is a more recent and fluid concept open to contestation. It involves interpretation of a legacy from the past and therefore requires the identification and valorisation of an authentic provenance. As we have established, this is commonly manifested in terms of buildings, monuments and the physical environment and artefacts and occurs through individual and group collective memory. Who controls this preservation and valuation process and what relationship such heritage has to the city – spatially, culturally and symbolically – are of increasing concern and debate, since the commodification of heritage assets creates economic benefits that accrue to property interests and the heritage tourism industry (Hewison 1987; Park 2013). Heritage has therefore moved from a benign, specialist concern to a central role in city branding, and the promotion of the city to its citizenry and to the outside world.

The importance of visible clues that anchor the development of cities to the past typifies the current desire to reconcile modern development and change with remnants of the city's past. This also reflects the wider democratisation of social history and urban archaeology – that is, the heritage of ordinary citizens and the everyday, for example, domestic houses, workplaces and leisure pursuits. Industrial and twentieth-century heritage itself is now subject to the preservation and value judgements applied previously to the 'historic', with the heritage question and heritage city branding increasingly applied to a wider range of locations, including urban heritage sites (Evans 2002).

World Heritage Cities

World Heritage site inscription was introduced following the UNESCO Convention on World Heritage in 1972. The Convention responded to the growing conservation and preservation movement and a rising concern with the deleterious encroachment of modernisation and modern construction and which, directly and indirectly, was responsible for the destruction of historic buildings, structures and views. World Heritage Site designation was thus an international recognition of the universal value and importance of a site. It brought both international branding (through the use of the UNESCO logo) and the installation of heritage management and interpretation measures to protect and control access to these sites and their surrounding buffer zones, which both seek to protect and further extend exclusion into these sites. In 2022, there were 1,154 inscripted sites, of which 845 are cultural or 'man-made'. The 250 cities which host World Heritage Sites (including trans-boundary sites) are members of the Organization of World Heritage Cities (OWHC). Over half of them are based in Europe; however, greater recognition and hard branding (Evans 2003a) of UNESCO World Heritage have seen inscription of city sites in the geographic and symbolic South, including the Stone Town (Zanzibar); Alhambra, Granada (Spain); Jerusalem (proposed by Jordan in 1981); Kyoto (Japan) and Oaxaca (Mexico). Heritage cities from past civilisations are not included however, since they are no longer populated or functioning settlements, for example the pre-Colombian Mayan cities of Central America.

The dominance of the Western European cultural hegemony in heritage designation has effectively exported this preservation ethic and system to other countries, mostly under the auspices of international heritage agencies such as UNESCO and ICOMOS in Paris. The preservation of heritage in towns and cities in non-European countries is a reflection of this movement, and also the imperative for heritage tourism and development aid from the West, notably the World Bank and international foundations (e.g. Aga Khan, Getty – Evans 2002). Conversely, the belated recognition of 'oriental culture' as seen from the perspective of western countries (Said 1994) now forms part of cosmopolitan city heritage in designated ethnic quarters and institutions, for example *Arab Monde*, Paris and Native American museums in New York and Washington DC. (Evans and Foord 2006a).

Living in a World Heritage City

Who 'owns' and mediates this heritage, culturally as well as legally, is often contested when political and cultural power shifts. For example, Quebec City, the administrative and political capital of francophone Quebec Province (and Canada), was founded (*sic*) by Samuel de Champlain in 1608. Its historic quarter, fortifications and battlefield sires were designated a World Heritage Site in 1985. However, the site is actually the British garrison which usurped the French garrison in the 1750s, won with the support of Native Americans, earlier explorers, traders and migrants. This particular contested or 'dissonant heritage' community (Tunbridge and Ashworth 1996) therefore represents local (residents), provincial, separatist, 'national' (Canadian), First Nation, migrant and colonial (English, French and Irish) interests and histories – but not on equal terms (Evans 2004). Universal recognition also presupposes that a case for national heritage status and significance has been established, and this site had been designated a national monument in 1952. As Wallerstein maintains, the 'national' completes the dialectic between the universal: 'in our modern world-system, nationalism is the quintessential particularism, the one with the widest appeal, the longest staying-power, the most political clout, and the heaviest armaments in its support' (1991, 92).

The protection and selection of heritage sites and monuments generally involve an internal process and campaigning of interest groups, notably conservationists driven by a reaction to, or threat from modern development and encroachment on sites and buildings themselves, including social displacement and environmental impacts. In urban areas in particular, heritage protection is also a response to progress and modernisation which threatens the representations of the past. In this sense, heritage designation is not just a benign exercise of securing an award for international status for posterity and humanity but a symbol of previous societies and cultures in situ, in the form of a 'museum without walls' and indeed some heritage sites are referred to as 'open museums', for example, Ironbridge, Shropshire, in the UK and 'living museums' such as Beamish in Northumbria, UK, and Williamsburg, USA. Malraux coined this term in 1967, claiming that the museum had been transformed from representing the world through universality and spectacle and later to order the world into a discourse of progress evidenced by the material residue of history, to effectively breaking the bounds of the institution by inhabiting public space (in Shaw and Macleod 2000). Hetherington also suggests, in the case of Stonehenge, that as a museum without walls, this site has become a heterotopia – a site used by many different ways by various social groups with both a legitimate usage and variety of alternative readings: 'the spatial trajectory in Malraux's account is one of ever more openness, from the private ownership of works of art, to private collections, to semi-public collections, to collections open to the public, to the final spilling out of cultural works into a generalised public over which the gatekeepers of the museum have less and less control' (1996, 155).

Old Quebec

The globalisation of populated tourist space is also a phenomenon increasingly evident in major cities, historic towns and urban heritage sites, which Edensor has identified 'not only as a recent development, but one which can be understood as an expansion of inscribing power through the materialisation of bourgeois ideologies since the nineteenth century' (1998, 11). Accordingly, places are now conceived not as nuclei of cultural belonging, 'foci of attachment or concern', but as 'bundles of social and economic opportunity competing against one another in the open market for a share of the capital investment cake' (Kearns and Philo 1993, 12). The reality of this is evident from the perspective of a Quebec City resident:

> In 1970 there were 59 bars and restaurants within the walls; ten years later there were 80. To this very day, Old Quebec has been undergoing a trans-formation into space for selling rather than space for living. The process of evicting a portion of the population beyond the walls whereas 20% of people who worked downtown lived there, the current figure is 11%. Old Quebec and its immediate vicinity is assaulted during the day by a flood of people who shuttle to and from work during the day, and during the evening and night-time by partygoers from the suburbs and the outskirts, and 24 hours a day by tourists and conventioners.
>
> (Evans 2004, 119)

Edensor also maintains that the 'processes of blending culture and capital have become increasingly disembedded from localities and become the provenance of an international class, who displace the paternalistic control exercise by local agents' (ibid.). Most importantly in this discourse however, is the fact that many visitors to heritage sites are 'domestic', that is 'nationals' (Prentice 1993) – whether the Great Wall of China, or Old Quebec where nearly 40% of visitors to the historic national site (HNS) are in fact Quebecois. Furthermore, local 'everyday' usage of sites often goes unrecorded – including local ceremonies, festivals and closed, off-season events. The preservation and presentation of heritage in this respect can therefore arguably be seen as a public good first; a celebration of national culture and glory to outsiders second; and given certain criteria, of universal value and patrimony – and therefore deemed worthy of world heritage site status. A dialectic between the universal and national, cosmopolitan and local, ignores the possibility that local and visitor experience of and relationship to heritage is complex and even contradictory, but one that is certainly not a binary divide.

　Whilst the national serves as a convenient if no longer robust (not least in Canada) counterbalance to the more abstract universal, the local in both the spatial and social sense also provides the most visible manifestation of host–guest interaction in practice (Smith 1997). The consideration of inhabitants or residents of course represents a particular temporal state, one which is likely to ignore previous and displaced groups to whom aspects of heritage may attach. In Quebec, this is clearly so in the case of the displaced and dislocated First Nation Iroquois, the merchant and military settlements from France and Britain and, in recent times, the communities displaced by urbanisation then suburbanisation, followed by gentrification and commodification of the historic core.

In Quebec city, the role of local residents which had been largely ignored and in decline both numerically and in economic power was reasserted due to the return of a small but articulate residents group in what has been a form of gentrification of the upper old town. This group also occupy the site of Quebec's political and symbolic heart, literally in the middle of competing nationalisms – Quebecois and Canadian, and related multi-cultural and pluralist debates (Bauer 2000). The incumbent residents, organised through the Old Quebec Citizens Committee, also have an interest in the conservation and promotion of the historic quarter and world heritage site, but not as a statutory body, but one literally cheek by jowl with the landscape and fortifications which represent the WHS itself, and the visitors who bypass their homes and the coaches who choke their streets (Evans 2003b). Resident's properties are also part of the historic site and their maintenance is therefore a responsibility which owneroccupiers bear. Their relationship with both the conserved historic area and tourists is also ambiguous, as a survey of over 700 residents revealed (Table 4.1).

Whilst the 'historic value' and 'pride' in the built and natural heritage rank highly among residents, the negative impact from tourism reduces these benefits, including the touristification and seasonality of local services (e.g. shops), and for a significant minority, indifference to world heritage site status is evident. The 'historic value' is an attraction in itself, but this does not translate in world heritage terms, where the 'international character' from visitors, acculturation, cosmopolitan city, rather than the symbolic significance (e.g. francophone military past and capital city of *New France*), is more important to these residents as a result of WHS

Table 4.1 Survey of Residents in Old Quebec

Major Attractions of Life in Old Quebec	%	Disadvantages of Life in Old Quebec	%	What It Means to Live in a World Heritage City	%
Historic value	20%	Parking	30%	Pride	44%
Beauty of site	18%	Lack of services	14%	Indifference	24%
Proximity of services	11%	Noise	13%	International character	11%
Atmosphere	8%	Traffic	10%	Economic benefits	5%
Proximity to St. Lawrence River	8%	Tourism	7%	Privileges	5%
Restaurants	8%	Lack of green space	7%	Recognition of beauty	3%
Lively neighbourhood	5%	Costs too high	4%	Special character	2%
Culture	5%	Presence of undesirable elements	4%	Historic value	2%

Note: n = 720.

status. This is both at odds with the politicised adoption of heritage in this case and the perceived benefits which world heritage site status confers. This status is used in a low-key way in the destination marketing of the regional tourism bureau and some tour operators but is a largely benign factor in the image presented for Old Quebec by its 'keepers', Parks Canada and other agencies (the WHS logo does not feature on federal signs or information boards). In a study of tourism promoters and operators in the city, Marcotte and Bourdeau (2002) concluded that world heritage listing is not generally a brand which is used to sell Old Quebec, and this association was only recognised by a small, highly educated and mainly European clientele (Evans 2002), but neither by the mass visitor market, notably Americans and Canadians, nor by local people.

Few cities, including Quebec, actually use the designation 'heritage city', since this generally applies to specific buildings, sites or areas, as in Lyons France, Bergen Norway, and Bath UK – within the larger city. The label can also refer to a collection of cultural and historic elements, such as museums and galleries, monuments, historic buildings and palaces or to a socio-cultural legacy as represented by living culture: notably, language, food, fashion, festivals, etc. Different aspects of heritage serve resident, visitor and tourist markets, particularly in terms of the historic, culture, nightlife and shopping attractions of the *Tourist-Heritage city* (Ashworth and Tunbridge 2001). The Heritage City, at the same time an urban imaginary and 'scalar narrative' (Gonzales 2006), is therefore one example of competitive-city placemaking, along with designations such as City or Capital of Culture, Knowledge City, Creative City, Science City and Sport City, among others (Evans 2009a). These are not exclusive, as cities strive to maintain multiple images and brands (and their different target markets). The absence of heritage in a city on the other hand implies a lack of past legacy of value and a dearth of opportunities for heritage tourism and self-identity, which only the most insular and autocratic city-state could risk.

Old world cities have, of course, had longer to reconcile the largely incremental pressures from growth and new building with preservation and heritage, whilst in new and resurgent cities such as Seoul, Shanghai, Beijing, Dubai and Kuala Lumpur, the heritage protection and value system are weaker – the 'new' is reified over the 'old'. In cities experiencing a resurgence of national and political freedom, legacies from the past can sit uncomfortably with both painful memories and new directions, as in former Eastern bloc capitals where communist monuments for example of Marx or Lenin, are dismantled or where despotic leaders are deposed as in Saddam City (Baghdad), first renamed Al Thawra 'Revolution City', then 'Sadr City' after the late Imam. The dismantling of the Palast der Republik ('People's Palace') in the former East Berlin and the recreation of the historic castle which predated it are an example of city heritage rejection and reversion. The redevelopment of the *MuseumQuartier* from the Imperial Stables in Vienna into a *Shopping Mall for Culture* is another version (Evans 2001, 219). Here, contemporary culture (museums, galleries and theatre spaces) was literally built within the walls, or façade, of this historic structure.

The globalisation of cultural space has thus fuelled the exploitation of city heritage sites as places of internal and external consumption. This latter phenomenon has become common to major cities and urban heritage sites. Historic City quarters along with medieval, old town areas have been 'rediscovered' and zoned as heritage assets, necessitating the displacement of poorer, often working-class and migrant communities. These now serve as locations for urban regeneration, with the siting of contemporary cultural facilities such as Pompidou in Paris, and MACBA in Barcelona, and the associated gentrification through adjoining residential, office and retail property use. Recent modern heritage listings include Oscar Neimeyer's Brasilia, Bauhaus sites in Germany, The White City of Tel-Aviv, the Maritime Merchant City, Liverpool, and Le Havre, France, whilst the addition of modern and industrial heritage include Chatham Dockyard in England and the Art Deco buildings of Miami. These listings reflect the growing intervention of architectural movements such as DOCOMOMO, and the recognition of modern architecture as worthy of preservation.

As the location for the ultimate accumulation of artefacts and architecture and subject to continuous interpretation, these cities stubbornly defy a single heritage branding such as 'Gaudi Barcelona' or 'Macintosh Glasgow' despite tourist agency marketing efforts. Moreover, the imperative for heritage policies and selection can be counter-productive, when the drive for being different and unique indicates how much cities have become the same. As post-industrial cosmopolitan cities multiply in number and in the range and layers of their heritage, they become both invisible and self-consciously visible through the official narratives and interpretation of heritage. As a consequence, cities become more and more alike (Calvino 1979). Entire cities do however adopt the heritage title where built environment and heritage legacy are of sufficient scale and homogeneity, notably in smaller cities such as Segovia, Venice and Florence. The use of heritage designation for industrial sites and cities has also provided a partial saviour for post-industrial decline such as in Lowell, USA, and Bradford, UK. Redundant industrial buildings increasingly serve as atmospheric sites for cultural facilities, whether celebrating industrial heritage itself (e.g. brewery buildings: Heineken, Amsterdam; Guinness World, Dublin), or more often, undergoing conversion for modern galleries and museums, such as Tate Modern, a former turbine station in London; Salts textiles Mill, Bradford; Musee D'Orsay, a former railway station and Parc de La Villette, a former slaughterhouse/market, in Paris and Venice's Arsenale maritime dock. Former industrial complexes in ports and docklands, mines, mills and manufacturing plants have also been recognised by heritage listing, such as Essen, Germany (coal mining), and the open museum at Ironbridge, Shropshire (UK).

A Tale of Two Ironbridges

Ironbridge is fêted as the birthplace of the industrial revolution, and its famous bridge structure was constructed in 1779, the first to be made of iron. This world heritage site was recognised in 1986 not just for its tangible and natural heritage – river, gorge – but also for its intangible legacy of iron, brick and wood crafting, making it a 'perfect destination for tourists and locals alike' (www.ironbridge.org. uk). Twenty years later in Spanish Town, near Kingston, Jamaica, a similar cast iron bridge was constructed in this suburb of Kingston, imported from West Yorkshire and bridging the Rio Cobre. Although never acquiring World Heritage Site status, in 1998, the World Monuments Fund (WMF) placed the Old Iron Bridge on the Watch of endangered sites because of its historical significance and critical condition. WMF provided a small grant for emergency stabilisation and long-term conservation of the site. A committee was formed by the Jamaican National Heritage Trust (JNHT) to designate an alternate route for traffic and to conceive a plan for repairs. They proposed to reconstruct the collapsed brick arch and stone-work of the north abutment, sandblast and paint the structure, and replace missing handrails. Unfortunately, progress at the bridge was stalled due to conflict in the region. The activity of feuding political gangs closed down businesses and gener-ally impeded life in nearby Spanish Town. For a period of time, the JNHT project coordinators could not find a contractor willing to work in such a dangerous area, and when Hurricane Ivan hit in 2004, the bridge was damaged even further, and work has not advanced much in recent years. Currently, the Old Iron Bridge is still being used despite a large hole and warnings from the authorities (it provides a shorter access route for local residents), adding urgency to the conservation project. The ongoing efforts of local organisations, despite violence in the area, are an indication of both the tenacity of the community and the importance of this troubled heritage asset. Once everyday sites of production, public amenity and flows (Lefebvre 1991), the historic significance of these two original examples of iron bridge design and construction, albeit in contrasting socio-economic con-texts and locales, are very similar. However, their value, valorisation and treat-ment by national and international heritage bodies could not be more different. Given their common provenance, the absence of any reference to the Ironbridge in Spanish Town in its Shropshire cousin is both telling and reinforcing the cultural hegemony at play in heritage preservation and promotion. In Spanish Town, this ruin has become in Simmel's words: 'a site of life from which life has departed' (1958, 384–385).

Heritage and Sense of Place

Given the ubiquity and exaggerated nature of participation in designated heritage in official narratives, research has also focused on attitudes towards heritage and historic environments. As well as actual visitor engagement for example, a signifi-cant number of people donate to the heritage sector (13% of those surveyed) and over 500,000 volunteer regularly in historic environments and organisations (His-toric England 2015). Membership of conservation societies has grown particularly since the 1980s in the UK, with over five members per 1,000 population, but with between 10 and 20 members in the London Home Counties. Attitudinal surveys also confirm that heritage – primarily the built historic environment – is valued for its contribution to our knowledge and sense of identity and because it helps to make a place feel 'special', with evidence emerging of a positive relationship between heritage participation, well-being and health (ibid.). Whilst conservation and campaigns to retain/renovate are generally focused on

> some aspect of the built environment which is taken to stand for or represent the locality in question, it is not merely that the object is historical, but that the object signifies the place and that if the object were to be demolished or substantially changed then that would signify a threat to the place itself.
>
> (Urry 1995, 156)

In a MORI survey conducted by English Heritage (EH 2000), over 90% of respondents felt that it was important to keep historic features when trying to improve their town, with concern about the state of buildings often acting as a motivation for people to take much greater interest in their local heritage (EH 2003). In this attitude survey, 96% of people thought that heritage is important to educate adults and children about the past and that all schoolchildren should be given the opportunity to find out more about England's heritage; 88% thought that it was right that there should be public funding to preserve the heritage; and 76% agreed that their lives are enriched by the heritage. Impacts arising from participation in heritage and public perceptions and attitudes can therefore be distinguished at three levels (Historic England 2014):

1 Individual (pleasure, fulfilment, meaning and identity, health and well-being)
2 Community (social capital, cohesion and citizenship, shared sense of place, civic pride)
3 Economic (Job creation, tourism).

The idea of *community* also presents an equally catholic range of interpretations, no less so than *heritage* – from Hillery (1955) to Urry (1995) who expanded Bell and Newby's (1976) analysis of this concept to four uses: first, the idea of community as belonging to a specific *topographical location*; second, as defining a particular *local social system*; third, in terms of *feeling of communitas or togetherness*; and fourth, as an *ideology*, often hiding the power relations which inevitably underlie communities. The values attached by people to what might be termed the historic environment will therefore be multiple, changeable and will not necessarily map onto those identified by official bodies. The historic environment should also be understood as a setting for people's daily lives, giving rise to a less conscious experience of place (CURDS 2009). In an international survey of *Creative Spaces* (Evans 2017) for example, smaller cities prioritised their heritage assets – built, natural, crafts, ethnic diversity and related tourism – as alternatives to creative clusters and pursuit of the creative class favoured by larger cities (Florida 2002). Smaller cities, particularly those with an historic or heritage reputation, have also resisted strategies to change the image and cultural profile of their city. For instance in Bruges, Belgium, European Capital of Culture (ECoC) in 2002, a clear conflict emerged between resident and tourist identification with its historic character – the very reason that ECoC event status was granted in the first place – and the bid organiser's motivation to promote a contemporary city (Boyko 2003). More recently, the French Marseilles-Provence region unsuccessfully twinned bucolic Aix-en-Provence and over 90 surrounding districts with its cosmopolitan port city neighbour, Marseilles in ECoC 2014 (Andres and Gresillon 2014).

The official evidence on economic valuation of heritage, on the other hand, tends to employ formulaic impact multipliers to estimate direct and indirect jobs, spending and growth attributed to heritage-related activity and the associated visitor economy. Proxy values are also imputed for free and non-market priced heritage usage and benefits, including transport cost models, contingent valuation (CV) and willingness to pay (WTP) or hypothetical 'stated preference' methods (Simetrica 2021a, 2021b), whilst 'revealed preference' seeks to quantify benefits through comparative property values in heritage areas, or in parks/green spaces. A controlled study of conservation areas for example found a 9% premium for properties inside such areas, an uplift that doubles in the centre of the area, and diminishing towards the edge (Colliers International 2011). A downside of the property uplift associated with historic buildings has inevitably been the loss of accessible heritage assets to private development, notably conversions for housing (e.g. church, school and municipal buildings), so that whilst facades may be retained, their future use value is lost. For example, Lister Mills (Grade II* listed) in Manningham, Bradford West Yorkshire, once the largest silk mill in the world, and once mooted as a *V&A of the North* – now developed for private residencies by developer Urban Splash.

Outside of the mixed economy of property and visitor economies, valuing heritage benefits has therefore relied on highly contingent valuation methods which have been developed as part of the cost–benefit analysis (CBA) of public investment and welfare goods and services, including culture (DCMS 2022). However, their validity is questionable with significant variations between hypothetical and actual willingness to pay (Kanya et al. 2019), and the results from these tests can be variable, even for the same individual over time and place (e.g. home, away, abroad), and between individuals/groups where affordability and value will also vary between the well and less well-off. Relying on these methods to scale-up quantitative values for heritage and cultural goods and apply 'benefit transfer' values elsewhere is therefore questionable. Heritage value studies are also limited to historic and natural environments, rather than locally determined and intangible heritage, and it is perhaps here that the concepts and practice of placeshaping and cultural mapping (see Chapter 9) provide deeper insights into how cultural heritage is valued, in all its forms (Evans 2022; TBR, Evans and NEF 2016).

Heritage Re-Use and Renewal

Heritage building re-use which recognises local knowledge and traditions starts from the proposition that getting at the meaning of places should not reside with professionals alone, but with the people who use/have used and who construct their own meanings out of places and buildings (Bazelman 2014). As Bluestone observes:

> [W]e need a system for taking measure of and working with the reception side of cultural heritage, conservators can take an active role, however they also need to be open to the possibility that the places they conserve for the purpose may take on very different meanings over time.
>
> (2000, 11)

However, the heritage adaptation phenomenon has typically been considered from an architectural/conservation (aesthetic/functional) perspective and from the imperative of economic viability and exploitation. Surprisingly, a cultural dimension has been missing from this field – one that considers the historic foundation, evolution and cultural values and identities attaching to these spaces of the civic past; their symbolic and innovative qualities; and how these might be transformed into sustainable re-use for a wider community benefit. This recognises that re-use too often looks to a reductive conversion option, for example as exclusive hotels, corporate offices and private residencies, that both excludes many residents/families, especially from the public realm and communal spaces, and tends to under-value their historic significance, bypassing the opportunity for productive and community use through cultural and creative activity, including start-ups, micro- and social enterprises (Evans 2016c, 39–40).

The re-use of heritage spaces for cultural activity has been a particular feature of regeneration schemes and local campaigns to save community assets. From the 1960s, arts centres located in former town halls, factories and industrial buildings – over two-thirds of centres originated this way, preserving key local historic assets (Evans 2001 – see Chapter 2), and across Europe, the re-use of large industrial complexes for cultural use *Trans Europe Halles* (Bordage 2002) has extended their life, including those arising from European Capital of Culture (ECoC) festivals, for example in the industrial heartland of Essen, Germany (ECoC in 2010). Local heritage assets have also been occupied by cultural and creative industries organisations and firms either during the 'meanwhile' period before redevelopment or as part of a permanent solution, with significant longevity, for example, the Omnibus workshops in Islington, north London, combining workspaces, studios, restaurants and the Pleasance theatre since the 1980s.

The clustering of creative industries in heritage quarters in former industrial districts is a now familiar scenario, from Beijing's 798 Art District, Sheffield's Cultural (renamed Creative) Industries Quarter, to the *MuseumsQuartier* in Vienna (Evans 2004, 2009b; Chen, Judd and Hawken 2015 – and discussed further in Chapter 5). These re-occupied heritage spaces can also attract a premium so that over time, more affordable arts and cultural uses can be priced out by higher-value creative firms such as IT, architecture and design studios as experienced in Clerkenwell Workshops, London, and SOMA, San Francisco (Evans 2009b). This scenario has been played out around affordable artist studios occupying former factory and municipal buildings, as well as creative industries quarters where gentrification and place branding attract larger firms (Evans 2014a). In order to mitigate this, some cities have either used specific planning powers to zone areas for (creative) industrial use as in Barcelona's Poblenou district or limited the size of workspaces to prevent larger conversions (or to retain larger footplates), as in Copenhagen (Evans 2009b). Even in these cases, however, creative industry gentrification has displaced incumbent arts and cultural enterprises, effectively crowding out established small cultural firms by higher value creative industries, as these zones become attractive and city government and institutions (e.g. universities/R&D agencies) extend their reach and impact.

The practice of place promotion has perhaps inevitably adopted the idea of place branding, drawing as it does from product and city branding strategies (Dinnie 2004). This practice has evolved from earlier place promotion and boosterism – the *art of selling places*, often in response to economic and social change and greater inter-place competition using heritage and historic associations (Ashworth and Voogd 1990). Place branding is a now familiar tool in city and local authority promotion, whether for tourism, inward investment or location decision-makers, including event hosting. Burgess's (1982) seminal study on the content of local authority promoted images in the UK-identified four main elements, and these are still relevant today: centrality, dynamism, identity and quality of life, whilst content analysis of the text and illustrations of 16 medium-sized cities in the Netherlands arising from Voogd and van de Wijk's 1980s study (Ashworth and Kavaratzis 2011) revealed the unanticipated conclusion that historical elements were being used widely in campaigns designed principally to attract external investment. Official brochures and printed advertising material (in a pre-website era) stressed historical events and personalities associated with the place. However, place branding as derived from product and marketing strategies is a limited concept when used alone, and is potentially risky, since brand ('product') decay is a familiar scenario, requiring periodic reinvestment and reinvention, whilst externalities or negative events can derail the image of a place. In the context of cities, particularly smaller cities and towns seeking to develop cultural programmes and event strategies drawing on local heritage, reconciling authenticity with the need to be both competitive and attractive for cultural and creative industries requires an appreciation of heritage from all of these perspectives (Figure 4.1), including how cultural heritage underpins place values and economy (City of Toronto 2010; Evans 2022).

As Arnold concludes:

> Another way to think about and approach the sustainable re-use of heritage buildings and sites is through co-design. This combines 'design thinking' that is to say creative and innovative processes we associate with the production of architectural and urban space with a more socially inclusive and iterative process of examining re-use options and scenarios that draws on historic and community knowledge and culture. In this way, we are able to explore the pros and cons of inscribing a building with a new use value set against social negotiations at local, national or global level and tensions between the use value and the worth of cultural heritage
>
> (2016, 183 and see Getty 1999).

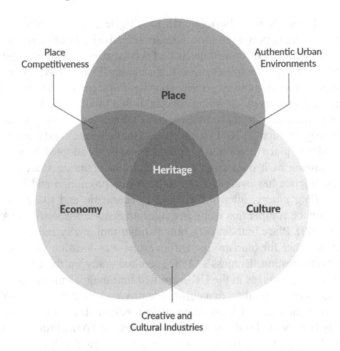

Figure 4.1 Place-Economy-Culture

The relationship between Place and Heritage, and Culture more widely, is long established – a 'spirit of place' or Genius Loci is defined by the traditions, legacies and everyday practices that make a community of place distinct and define its identity, manifested also through historic buildings/sites, artefacts and landscapes. Place, or rather a sense of place, is therefore used to explore different aspects – including topographical, built, environmental and people's own experiences – which together makeup the character or local distinctiveness of a specific place. A sense of place is also used to emphasise the way in which people experience, use and understand place, leading to concepts such as 'place identity', 'place attachment' and 'place dependency' (CURDS 2009). The contemporary practice – or rediscovered practice – of placeshaping therefore seeks to identify, enhance and better communicate these assets and senses across these physical, symbolic and everyday dimensions. Richards (2017, 5) argues that the triad of physical, imagined and lived space (Soja 1996) can be mapped onto a social practice framework (Shove, Pantzar and Watson 2012), where imagined space gives and takes meaning from the social and cultural context, and the everyday lived space is the recursive result of the creativity and knowledge of users. Through this practice, the interplay of these three elements helps to shape heritage and place (Table 4.2).

A fuller critique of placeshaping and how culture contributes to this is available via the DCMS CASE (evidence-based) study programme (Evans, NEF

Table 4.2 Relationship Between Place, Heritage and Placeshaping

Elements of Place	Placeshaping as Practice
Tangible Heritage *Physical space*	Buildings/Materials, Morphology, City/townscape, Open/ Public space
Intangible Heritage *Imagined/Symbolic space*	Meaning – traditions, symbols, characters, identity, memories, oral (word, song)
Living Heritage *Everyday Experience*	Creative engagement, local festivals, local history societies,

and TBR 2016; Evans and TBR 2010). Led by the author, data scientists and economists, these studies concluded that both the concentration of creative firms, measured by location quotients (LQs) and turnover, were positively and significantly associated with the density of heritage assets and the number of cultural events listings (measured per capita). A study for Historic England also surveyed over 60 Business Improvement Districts (BIDs) in England and found a positive association between heritage assets and narratives and business location and growth (Evans 2016), with heritage images frequently used in BID promotional material.

Although *Place* has surprisingly not featured explicitly in heritage policy until recently – and as already noted, place is largely absent from cultural activity surveys – English Heritage (EH) was one of a number of agencies which sought to demonstrate the importance of place to wider social outcomes. *Power of Place* stemmed from the belief that 'the historic environment has the potential to strengthen the sense of community and provide a solid basis for neighbourhood renewal' (EH 2000, 23). A related term also featured in more recent cultural policy: 'we want to see more partnerships being formed between the national and local levels to put culture at the heart of placemaking' (DCMS 2016, 29). The term placemaking however is normally associated with urban design and public space improvements, but this can be problematic, inferring that a place needs a makeover and that a place can be 'made', rather than building on its endogenous heritage past and present which may well be hidden or undervalued. In some cases, creative and sensitive placemaking can help to improve the reputation of a place from both resident and visitor's perspectives, but only where part of culture-led regeneration (Evans 2005) and co-created placemaking (Markusen and Gadwa 2010).

Public art schemes and street art (Evans 2017, and see Chapter 8) are also increasingly adopted in this way, the most successful celebrating an area's cultural and industrial heritage, for example, Anthony Gormley's *Angel of the North*, Gateshead, and a recent 82-metre-high mural in Leicester referencing logos and mascots from city football and rugby clubs, the National Space Centre, and DNA fingerprint research from nearby Leicester University. It should also be acknowledged that initial reaction to public art can be negative and subject to resistance (Hall and Smith 2005). When sculptor Antony Gormley was selected as the winning artist in 1994, his designs originally caused uproar among locals and media alike. The controversial material and site of the sculpture were frowned upon; however, once in place many people's original views on the piece changed and it has become synonymous with nearby Gateshead. Heritage and its historical context are, in many senses, both subject to being rewritten and rejected, long after its first installation, most recently experienced with statues of once-respected philanthropists and dignitaries who have been associated with slavery or colonisation. Public art need not, of course, be immutable, running the risk of closing off future use and alternative perspectives. The Fourth Plinth project has been one novel response to this challenge in an iconic site bounded by the National Galley and Portrait Gallery and several foreign embassy buildings (including Canada and South Africa), located in London's Trafalgar Square, a site of national heritage (Nelson's column), gathering and protest and flows of visitors and pigeons. The fourth plinth had remained empty for over 150 years when it was first intended to hold an equestrian statue of William IV, but never executed due to lack of funds. A competition for artist commissions has been held since 1998 with installations mounted for public viewing for two years, allowing passers-by and other commentators to experience and express their view against a backdrop of fixed monuments – for example, *The End* by Heather Phillipson (displayed between 2020 and 2022), a dollop of whipped cream topped with a cherry, a fly and a drone which filmed passers-by and displayed them on an attached screen (Figure 4.2).

In practice, a combination of different approaches is needed in order, as Zenker put it, to 'catch a city' or place (Zenker 2011). It is suggested that approaches encompassing cultural asset mapping and qualitative research on heritage associations and values can aid the general understanding, since identity exerts a strong influence over the perception of a place. For example, in a study of Mons' *European Capital of Culture* 2015, the concept of *memoryscape* was used to understand how the working-class memory of the city was managed during this cultural year (Basu 2013). Arts-led research in heritage settings incorporating co-design and participatory methods have included oral history, drama, dance and visual arts, multi-art-form festivals and residencies, as well as creative methods practiced in community planning and architecture, notably *Design Charrettes, Parish Maps* and *Planning for Real* (Evans 2008). These are discussed in more depth in Chapter 9, including case studies of socially engaged practice in the context of urban cultural heritage. Participatory visual mapping in particular (Evans 2015a) has proven to be helpful in identifying intangible elements and space–place relationships, as illustrated here in the case of an urban heritage building.

Sutton House, Hackney

In this example, a simple form of participatory GIS was used in an urban heritage setting in east London, as part of the UK National Trust's *Parent Power* pilot scheme, which was designed to improve understanding of what makes a historic house or museum a place to be enjoyed by families, especially those from disadvantaged and culturally diverse communities. The scheme worked with local

Figure 4.2 The End, Fourth Plinth, Trafalgar Square (viewed via The National Gallery)
Source: Photograph by the author.

families, the London Borough of Hackney and Hackney Museum Service, where parents and children were encouraged to explore and discover the range of opportunities available for families in this borough. This was felt to be important in both access and cultural policy terms, since ethnic minority groups have significantly lower participation rates in cultural facility usage, as discussed earlier. In discussion with National Trust staff, the case study was designed to investigate visitor movement and preferences around an urban historic museum, Sutton House – a listed building – in order to determine:

- How long they spend in every room?
- What they like in each room?
- What they remember about the spaces?

The National Trust had encouraged the managers of its historic houses to arrange family trails to help visitors maximise their enjoyment of exploring their properties, and in the case of Sutton House, information received from using this method was intended to be used to design such a trail. By experimental use of GIS-P at a micro-level to map the perceptions and enjoyment of the historic building on a room-to-room basis, it was anticipated that the process would inform the physical operation and resources available to families at Sutton House, especially those that had little commitment, to ensure that these initiatives would benefit local users as well as other visitors. By involving the selected families that were already participating in the *Parent Power* initiative, the case study provided practical means of furthering this work. In particular, it enabled the families to express their responses on large-scale paper plans of Sutton House (room plans rather than maps were used in the sessions) as illustrated in Figure 4.3, although this had the disadvantage of lacking a three-dimensional representation in terms of furnishings and other objects (Evans and Cinderby 2013).

*

As this example illustrates, cultural heritage – as selected, represented and interpreted through its often historic productive, symbolic and functional origins – still largely exists and is experienced in the everyday. These legacies are generally valued more by their inhabitants than official monuments and sites, likewise the vernacular aspects of designated heritage – including the intangible, and both are highlighted and regularly feature in user experience surveys. In our study of everyday heritage and business improvement districts for example (Historic England 2016), locating in heritage buildings and areas is seen to enhance both the identity and value ascribed to local firms, particularly smaller organisations. Users also identify with these past uses and re-use, irrespective of the current contemporary occupation, from bank to bakery and cinema to café. It is no surprise therefore that cultural and creative enterprises have chosen to locate and adapt heritage spaces for a wide range of production and related consumption activity, and this operates at the scale of districts and quarters, down to dense concentrations and urban transects of mixed-use premises (Evans 2014a). The phenomenon of cultural and creative quarters and the convergence of production and consumption are therefore explored further in the next chapter, including their exploitation as subjects of placemaking and as innovative cultural spaces in their own right.

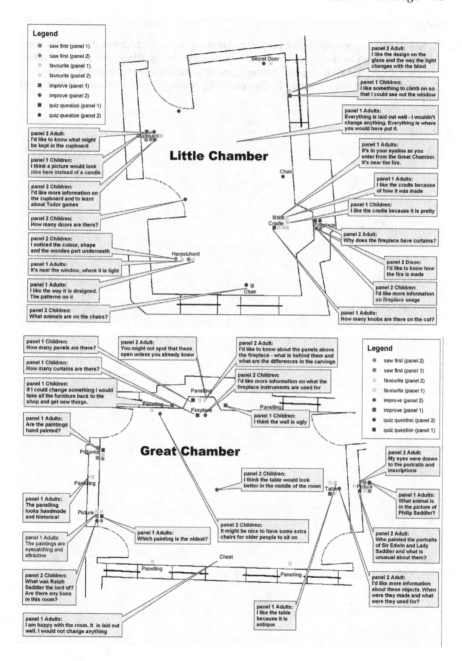

Figure 4.3 Sutton House GIS-Participation Comments

Notes

1 See AHRC Leadership Fellow project by Dr Susan Ashley, which sought to understand the ways that non-mainstream cultural organisations led by Black, and minority ethnic members engage with ideas of 'heritage', www.academia.edu/video/lDdoVl?email
2 Based on selected museums in France, Italy, Spain, The Netherlands, Greece and Poland.

5   Cultural and Creative Quarters

Spaces for arts and cultural activity are generally associated with facility-type provision, including larger complexes and clusters of civic and commercial institutions, notably museums and galleries, theatres and larger arts centres and educational facilities, for example, university campuses and conservatoires. Their connectivity and relationships can vary from minimal, stand-alone status to being part of a local and regional ecosystem, meeting a hierarchy of cultural development needs (Evans 2001, 2008). Cultural production, on the other hand, has from industrial times traditionally been separate(d) from consumption, both spatially and economically, as well as creatively, from individualised and small group creation (e.g. studios/workshops) to more industrial scales of production, reflecting the spatial separation of home, work and leisure from the industrial planning era. This has included concentrations within and across cultural production chains, in many respects mirroring the clustering familiar in other forms of industrial production. Indeed, cultural clusters have represented this phenomenon in pre-industrial eras, seen in crafts districts within and on the edge of towns and early cities, for example classical Athens (Evans 2001). Today, cultural production continues to exhibit this tendency, reflected in cultural and creative clusters in both traditional and new modes of production. Given their longevity and preference for location loyalty, this has entailed occupation of industrial districts and spaces that brings them into the remit of heritage preservation, promotion and, more recently, placemaking. This in turn has placed them at the centre of local economic development efforts and a reinvention of a role as places of cultural consumption, a coming together that neither Marx nor Lefebvre could have envisaged. Cultural producers in this sense have also become subjects of observation and interest, as well as generating their own scenes and lifestyle-related amenities in close proximity to their workplaces.

Cultural and creative quarters ranging from historic districts to new digital hubs have therefore featured increasingly in economic development and industrial cluster studies (Pratt 2004; Foord 2009) – from Shanghai (Gu and O'Connor 2014) to Singapore (Gwee 2009) – and with the emergence of creative placemaking (Markusen and Gadwa 2010), the identification of quarters has been a natural development in the conception and practice of place branding. Essentially production based, they have tended to be located in industrial areas of cities which are also the subject of or prospect for regeneration and transformation, which in many cases

DOI: 10.4324/9781003216537-5

represents a proxy for gentrification (Evans 2005). As cities seek to widen their cultural experience 'offer' and diversify the range of destinations to a discerning visitor and residential market – local, domestic and hopefully international – these areas, which increasingly combine work and play, represent a distinct assemblage of cultural production and nascent consumption spaces, and where 'social space has an affinity with the Deleuzean "assemblage" (Deleuze and Guattari 1987) or the broader "event-scene", even in its simultaneity it is the place that maintains primacy'.[1]

Place Branding and Making

Although commonly conflated and used interchangeably in policy and practice, place branding should be distinguished and considered separately from both placemaking and placeshaping (Evans 2014a, 2016c, 2022). Place branding is the archetypal process of conceived and perceived space production. City-branding strategies and the practice of placemaking have a wide range of rationales and effects. They may inter-act or more often represent quite separate processes and operate at different (macro and micro) spatial scales. The extent to which branding is a conscious and explicit goal of urban development is also variable, in part due to the essentially incremental nature of most land use and economic (as well as social and cultural) development and in part due to the limitations of branding as a concept and sustainable approach to city development and management (Evans 2006). The treatment of cities as commodities in need of marketing and selling themselves (Ward 1998) is commonly rationalised for example by Harvey's perceived shift from managerial to entrepreneurial urban governance (1989; Hubbard and Hall 1998) in response to the twin forces of post-industrialisation and globalisation, and the growth of a so-called network society (Castells 1996). However, as Tuan observed earlier, boosterism through cultural facilities has been long practiced – from pre- to post-industrial times – and it would be simplistic to paint city management as a free market process at the cost of local amenities and governance. As more pluralist regime theories suggest (Stoker and Mossberger 1994), the reality is more complex with quality of life, distributive equity and local economic development still mediating more commercial imperatives at city and regional levels, notwithstanding branding efforts which whilst occasionally high profile generally represent a marginal aspect of city planning and development outside of specific sites, events and regeneration zones.

For example, in an international study of creative city policies and strategies (Evans 2009c), branding ranked behind economic development/job creation, infrastructure and regeneration (Figure 5.1) as the prime rationales for public cultural investment and creative industries policies. In the case of smaller cities, with a population of under 200,000, often without the economic, political or cultural clout to engage in expensive and risky imagineering, *Branding* was the least frequent reason given for investing in creative city campaigns and strategies (Evans 2012b), the creative city meta-brand either being outside of credible reach or rejected altogether, along with Florida's notion of a footloose creative class (2003) which is

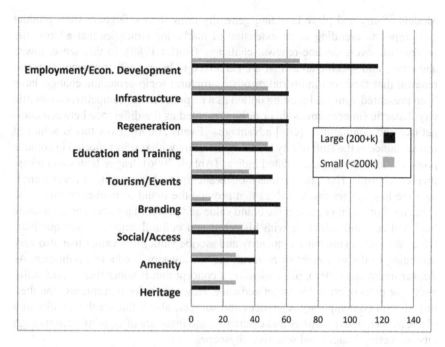

Figure 5.1 Creative City Policy Rationales in Large and Small Cities

seen by these smaller cities to be at odds with vernacular, diverse and indigenous culture and community creativity (Evans 2009d).

Whilst place *branding* infers a product-service experience, albeit an intangible prospect that is both ambiguous and mutable, place*making* is generally a case of physical intervention in urban space and the built environment, hence its appeal to planners and politicians alike. Lew for instance outlines four key types of place-making which usefully distinguishes their intent and potential impact, with notably different processes and outcomes (2017):

1 *Standard placemaking* focused on physical upkeep and maintenance of the built environment.
2 *Strategic placemaking* focused on the creation of a new development on the scale of a neighbourhood or city through a top down development approach with a significant level of investment, often from governments and/or private developers (Evans 2005).
3 *Creative placemaking* focused on the utilisation of the arts, to make a place more vibrant and interesting, be it through applications to the physical environment, the presence of arts related businesses, or the staging of programming and events (Markusen and Gadwa 2010).
4 *Tactical placemaking* focused on a 'bottom-up' approach led by community groups looking to test, change or improve aspects of their locale and often using temporary, low-technology/low-cost interventions (Evans 2022).

Models of city and place branding generally draw their references from product and corporate branding as an extension of marketing strategies that address the product life-cycle decline-renewal challenge (Butler 1980). In this sense, towns and cities, and specific areas that are perceived to be in need of regeneration and renewal that face post-industrial or other structural socio-economic change, have been presented with the branding option as a response to the competitive-authentic city dialectic (in economics this can be presented as the difference between competitive and comparative [city] advantage – Evans 2009b). How this is achieved and sustained is the stuff of city branding literature with results reflected in copious proprietorial branding and related indices (Anholt 2006), league tables and measurement formula. Here, the various models attempt to disaggregate or reverse engineer the key factors and variables that provide the brand or marketing mix – the elements that together present the brand value and power of a place. These combine hard and soft infrastructure with historical and cultural amenities and qualities, which themselves are hard to quantify and ascribe values to – values that also vary according to the viewpoint of resident, visitor, investor, media and politician. As Zenker maintains (2011), place *identity*, a concept that is wider than *brand*, influences the perception of the target audience. However, prior perceptions, and their historic and contemporary references and sources, also influence the identity of a place, as seen both internally and externally, and these are often reinforced through city marketing images and selective cityscapes.

A particular aspect of placemaking therefore is the landscape used and appropriated through branding and city-marketing strategies. In city branding models, the cityscape is characterised variously as *place physics* (Anholt 2006) and *spatial picture* – as distinct from amenities such as historical and cultural facilities – theatres, museums and parks (Grabow, Henckel and Hollbach-Grömig 1995; Grabow 1998). Kotler et al. (1999) prioritise design or *place as character*, distinct from 'infrastructure' and 'attractions' in their place improvement and marketing approach, whilst Ashworth and Voogd (1990) first proposed a *geographical mix* to capture the 'whole entity of place-products' (Kavaratzis 2005, 336). Infrastructure typically includes transport systems and facilities, hospitality and increasingly connectivity and digital highways. The latter is used as part of city promotional strategies in so-called Smart cities such as Seoul, South Korea, as well as in digital *free-wifi-here* zones. However, despite the physical imagery and changing skylines strongly associated with city and place branding and destination marketing, it is interesting to note that in Zenker's (2011) analysis of 18 place branding studies (2005–2010), architecture, buildings and city cultural spaces were largely absent in the brand elements cited. These surveys tended to focus on generalised or experiential associations cited by residents and visitors such as *culture, historical and buzz*, rather than specific physical attributes. It would be unlikely if a brand awareness study of Bilbao ranked 'art museums and galleries' as its key element (it has several), whereas Gehry's Guggenheim building would provide the key rationale to visit and in city brand awareness (Plaza 2015). Here, the pre-fix indicates the strength of a particular architectural city identity (and legacy), for instance *Gaudi Barcelona, Macintosh Glasgow*, along with the more recent (and thus less diversified) *Guggenheim Bilbao*.

For example, in the case of Guggenheim Bilbao, using firm location data, Plaza, Tironi and Haarich (2009) mapped the cluster of art-related establishments (art galleries, auction, antique houses, etc.) before and after the Gehry art museum franchise. Whilst this study found that galleries had relocated closer to the Guggenheim and new establishments also opened up in the same area, this was not at the cost of other cultural districts in the city. A separate – and more 'creative' – cluster of 32 contemporary art galleries, antique and auction houses has developed in the Casco Viejo area adjoining the Guggenheim district, including handicrafts, manufacturing as well as retailing and art market activity (TBR, Evans and NEF 2016, 20–21). However, this apparent micro-cluster gathering around the Guggenheim is neither economic nor cultural: 'The arrival of the GMB to the neighbourhood created positive (and somehow overrated) expectations of sales, and many art gallery owners relocated their businesses to Abando' (Plaza 2009, 1721) – see Figure 5.2.

In fact, these owners misunderstood the functioning of the art market: visitors to the area are sporadic, short-term travellers and, therefore, are not the target market of these businesses. The relationship of the art galleries and their customers, on the contrary, is based on long-term trust and on an exclusive and highly personalised treatment. In other words, many art gallery owners located in Abando wanted to take advantage of the Guggenheim's impact on tourism, when their core target

Figure 5.2 Art-Related Cluster in Inner-City Bilbao, 2006

Source: Plaza, Tironi and Haarich (2009, 1721).

was located in the traditional (and almost unaffected by the Guggenheim), local art market (ibid.). Physical clustering and co-location may therefore not represent either a creative or collaborative locality; it is the relationships – supply and production chains, shared objectives and social networks – which are characterised by:

- communities of creative people (*communities of practice*) who share an interest in novelty and new things (wherever they occur)
- a catalysing place – where ideas and connections are sparked
- diversity of experience and freedom of expression
- dense but open networks of personal relationships that permit identities and uniqueness to flourish (De Propis and Hypponen 2008).

These conditions and factors are not captured visually in a map or database but require a degree of understanding, presence and participation in these networks of shared space in order to gauge the nature of the location decisions and their relationship with place.

Cultural Quarters

Whilst the literature on iconic building projects is now extensive (cf. Ponzini and Nastasi 2011; Glendinning 2010, 2; Foster 2013; Evans 2001), their brand impact and significance are still under-developed and surprisingly less considered in city branding literature (Zenker and Beckmann 2013); however, they do represent one of the most visible manifestations of city branding (see Beatriz's Plaza extensive analyses of Bilbao Guggenheim's effect on the place brand, e.g. Plaza et al. 2015). Conversely, the less iconic urban design and placemaking associated with established and emerging cultural and creative quarters represent a new aspect of city branding that is more organic and arguably authentic, and more integrated with the post-industrial urban economy and therefore creative city aspirations (Evans 2009b, 2009c). The cultural or creative quarter can thus build on symbolic and heritage legacies and also create new destinations and experiences in areas that have not formerly been identified as places of interest to visitors, or be part of city brand portfolios.

A long-practiced concept in urban design and planning is therefore the spatial designation of *quarters* by dividing areas of the city into discrete and congruent zones that reflect their historic land use, morphology, economic and social mix. According to new urbanist, Krier:

> the urban quarter is a city-within-a-city, it contains the qualities and features of the whole, provides for all the periodic local urban functions within a limited piece of land, and are zoned block-wise, plot-wise or floor-wise. An urban quarter must have a centre and a well-defined readable limit.
>
> (1995, in Montgomery 2013)

They can build on pre-existing areas, notably historic or heritage, business/CBD or university districts, retail and entertainment zones. As one of the leading architect-masterplanners Rem Koolhaus claims however: 'progress,' identity, the city and the street are things of the past' (1995, cited in Glendinning 2010, 114). So in his elevated view, cities are either faced with the individualism and karaoke architecture (Evans 2003a) promoted by international firms and their booserist clients or a combination of an organic and a plan-led approach to urban design – a conflict perhaps between city branding and creative placemaking (Evans 2006). This former formula is not guaranteed to succeed however, where import of iconic structures has found resistance, informed by past failures and controversies, for example the cancelled *Cloud* building designed by Wil Alsop as Liverpool's *Fourth Grace* as part of the city's European City of Culture festival (Evans 2011); the aborted Cardiff Bay Opera House (Zaha Hadid); and Rio's rejection of a Guggenheim clone. Nonetheless, Alsop's Cloud design was mooted to be relocated to Toronto, one of this city's 'million dollar babies' – a sure sign of its inauthenticity (Evans 2009c).

Urban spatial interventions that have come to reflect aspirations which are adopted in city-branding portfolios can take several forms (Table 5.1, Evans 2014a). These are increasingly linked and are not therefore exclusive, with events and festivals spawning new structures and quarters of various types imposing branded design and wayfinding and legibility images through signage, street furniture, banners and other markers. Open spaces including parks and squares are frequently used as focal points for events and celebrations, sculpture and other art and media installations, in some cases a rediscovery of their pleasure garden past, although this can produce conflicts with their amenity value and usage (Smith et al. 2021). They can also anchor cultural quarters and entertainment zones, ranging from the spectacular, permanent to the everyday and ephemeral (Evans 2004).

Table 5.1 Forms of Urban Design and Quarters

Spatial Form	Key Types	Examples
Urban design/ Placemaking	Public squares/Pleins, Routes/avenues, Parks/ trails, Pedestrian zones, Public art, Wayfaring/ Wayfinding	Centenary square Birmingham, Park de la Villette Paris, High Line New York, Olympic Park London, Barcelona waterfront
Ethnic Quarters	Area/street-naming, signage, Gates, Street furniture, Festivals	Chinatowns/gates, Curry Miles, Banglatown London, Little Portugal Toronto
Heritage and Cultural Quarters	World Heritage/historic sites, Heritage quarters, Arts districts/Culture parks, Cultural Industry clusters	Saltaire Bradford, Lace Market Nottingham, Distillery District Toronto, Hua Shang Culture Park Taipei, 798 Art District Beijing
Creative Industry Quarters	Creative industry production zones, Managed workspaces, Incubators, Live-Work premises, Digital Media/ Techno parks	Digital Shoreditch London, Liberty Village Toronto, Amsterdam-Noord, Sheffield CIQ, Prenzlauer Berg & Kreuzberg Berlin, Creative Enterprise Zones (CEZ) London

Creative quarters, or clusters in the economic sense (i.e. agglomeration), can be seen not only as examples of mutual co-operation through informal and formal economies of scale, spreading risk in knowledge/skills/R&D and information sharing via socio-economic networks, but also as reactive anti-establishment action (e.g. avant-garde, artists' squats) and as a defensive necessity, resisting control from licensing authorities, global firms, guilds and dominant cultures – artistic and political. The factors that contribute to this concentration and proximity include cost-savings in the production chain, cross-trading, joint ventures – for example in marketing/showcasing, IT and capital investment – and the rediscovery of live-work arrangements – architectural, social and production (Holliss 2015), as well as shared workspaces within former industrial zones and buildings. Lifestyle and other synergies are also emerging as pull factors in clusters of firms in both traditional pre-industrial arts (Lacroix and Tremblay 1997, 52) and in new media services (Backlund and Sandberg 2002). These processes have come together in the regeneration of former industrial districts and buildings that served old crafts production, for example, textiles, ceramics, jewellery/metalcrafts, which, following manufacturing decline accelerated by offshore production, are being redeveloped for new creative industry quarters and hoped-for innovation districts.

The development of a high concentration of cultural workers and facilities for public consumption has also been a familiar aspect of theatre-lands and designated entertainment zones such as in the West End, London; Broadway, New York; Rio's cinema-land, Berlin and London's museum quarters, and Amsterdam's red light district (Burtenshaw, Bateman and Ashworth 1991), but this can also be seen in non-public cultural activity which focuses on production spatially separate from creation/distribution/dissemination, such as in Soho's film/media and music post-production facilities and in California's Silicon Valley (Scott 2000), both in their ways, seminal creative production spaces. Versions of agglomeration and cultural industry quarters can be seen (or not, i.e. they are hidden but none the less active), bringing together a range of compatible elements in the particular production chain, whether audio-visual, design, crafts, visual arts or producer services based. For example, the artisan villages in the Modena region of northern Italy have played an important part in the area's renaissance since the late 1970s and 1980s, through the flexible production of individual arts and crafts settlements made up of a wide variety of small manufacturers (Lane 1998, 158). These form a network in which companies are competitive with and complementary to one another, in common with small craft producers in managed workspaces (Evans 2004). These producer zones in turn form a polycentric grid throughout the region, which has ensured their competitiveness over manufacturers (e.g. furniture, textiles, ceramic tiles) in traditional 'chaotic' areas, such as in East London (Green 2001). As Scott maintains: 'The cultural economy of [late] capitalism now appears to be entering a new phase marked by increasingly high levels of product differentiation and polycentric production sites' (2001, 11). Where cultural producer clusters or industrial districts are long established, their survival and development have also reflected structural changes in production techniques and technology, as well as markets and cultural development in both design and consumption/fashion trends. This is evident in the profile of cultural production in traditional quarters where new media have replaced print and publishing (e.g. from magazine to web site); metalcrafts and weaving have evolved into multimedia jewellery and textiles production; and painting and sculpture are supplanted by media art and time-based film and digital media installation – a shift from (hand)craft artisan to designer-maker and producer.

The continuity has therefore been in the place and spaces occupied for this activity, rather than the precise forms of creative production themselves, although some residual continuity is still evident in the performing arts, metalcrafts/jewellery and specialist services such as instrument makers, costumiers – skills often passed down through the family, and which have not been so easily automated or 'designer-labelled'. Moreover, where established manufacturing activity requires updating and a more responsive mode to market and consumer demands, secondary or complementary clusters form, which are able to feed the traditional production district and filter design and innovation emanating from art schools and designers. This has occurred in New York (Rantisi 2002), London's east end and the Nord-Pas de Calais region of France (Vervaeke and Lefebvre 2002), which support a traditional sweatshop design and manufacturing district linked to major retailers, and an inter-dependent but culturally distinct new quarter served by art and designs schools, specialist boutiques and independent designers/makers, as discussed further in Chapter 7. In Nottingham's Lace Market (East Midlands, England), this fashion and textiles-oriented quarter is embedded in a conservation and regeneration zone in close proximity to the city centre (Crewe and Beaverstock 1998), but this has become more heritage-tourism (including speciality retail) than producer-based. In New York's Garment district, industrial premises have increasingly been turned into high-end retail shops, with loft living apartments above ground floor – a similar pattern seen in London's historic Clerkenwell district where light industrial buildings have been converted into higher (property) value residential and office use above, and expensive restaurants and showrooms at ground level (see Figure 5.6). However, some larger creative workspaces in former industrial/works buildings have survived this commercial gentrification effect, fuelled by more profitable advanced producer services in design and 'creative' business services, which are able to pay substantially higher rents (Evans 2004). These form part of a geographically wider creative production chain and are supposedly populated by Florida's expansive creative professional class (Florida 2003), or more accurately, 'knowledge-intensive workers' (Evans 2019), rather than artists/bohemians and crafts workers.

As well as a planning and development mechanism allowing for effective zoning of a city, quarters are also used as an area regeneration, conservation and economic development strategy. The latter includes destination and tourism management which encompasses area branding around heritage, entertainment or other visitor-oriented districts (Roodhouse 2010). As discussed in Chapter 4, Conservation Areas have been used since the 1960s to protect and preserve the built environment and, to a lesser extent, usage and occupation, with historic building and heritage listing or grading used by planners and government to preserve heritage assets. Urban heritage sites seeking and gaining listing are also increasing as the value of and imperative for conservation intensities. City branding therefore looks to the historic stock and heritage sites as key elements in visual imagery and brand associations. Heritage quarters are also the subject of further development, both to extend the quarter and modernise its facilities. For example, the extension of *The Lighthouse* building first designed by Rennie Macintosh in Glasgow for the Herald Newspaper (and now the Centre for Architecture, and hub of the 1999 *Festival of Architecture*); and the long term, if inauthentic 'completion' of Gaudi's *Sagra Familia* in Barcelona (including the recent addition of a five-ton glass and steel star atop the Virgin Mary tower). Reclaiming industrial heritage has thus benefited cultural institutions with not only newly-renovated iconic buildings which add to the available cityscape, but also urban amenity through regenerated open space and routes. For instance, New York's greening of the meat-packing freight route – the High Line – has created a newly accessible green route above the streets of the city. This has added a new attraction, an evolving pedestrian amenity for residents and a new placemaking imaginary easily connecting several parts of the city.

Ethnic Quarters

A particular cultural area with both symbolic and economic significance is the *Ethnic Quarter* – designating a residential/commercial/cultural neighbourhood through its association with a migrant community – from the ubiquitous Chinatown, Jewish Quarter (and 'Ghetto'), to *Little* – Italy (London, New York), Portugal (Toronto), Germany (Bradford) and so on. In many cases, the communities have long moved on (or were never there in the first place e.g. *Barrio Chinois*, Barcelona), but their heritage is manifested through signage, street furniture, place/road naming and banners and residual activities such as restaurants and souvenir shops. These are important for placemakers however, since they provide an eclectic offer and diversity that the city may otherwise lack, reinforced through annual community festivals and events (Chapter 3).

An extreme ethnic quarter makeover is seen in east London's rebranded *Banglatown* (Shaw 2012), centred on Brick Lane in an area that has experienced successive waves of migration and occupation over several centuries – from Methodists, Hugenots, Jews, to Bangladeshis and now East Europeans. Nearby this symbolic lane, an attempt to capture the multicultural city in physical form and place, recognising its absence and marginal position in the past, is seen in The *Rich Mix* cultural centre – a new-build arts centre designed by Penoyre and Prasad which aims to be a focal point for local communities, a meeting place, entertainment and educational centre (resonant of nineteenth-century *People's Palaces* and twentieth-century *Maisons de la Cultures/Arts Centres*). It seeks to 'challenge and strive for creative excellence over a range of art forms, a crucial crossroads, dedicated to innovation and integration, working towards a new understanding of British culture' (Evans and Foord 2006a, 74). What is being unsaid here is the multicultural basis for this venture, which is only manifested in its multiscreen cinema combining mainstream and Bollywood films and home to a music training agency initiated by the band, *Asian Dub Foundation*. The centre's location (and funding) seeks to play a major role in the regeneration of its local area – an area that has been subjected to office and residential gentrification and development prior to the centre's formation. This optimistic cultural development was based on creative city principles (Landry 2000), focusing almost exclusively on creative industries and related retail, hospitality (curry and balti houses, wine bars, designer retail and galleries, etc.) and street markets. The multicultural residential neighbourhoods have largely been neglected by this consumption-led approach, creating a socio-spatial divide with social programmes which promoted training in new media and patronising capacity building, but which ignored the local meaning and memory of place and the cultural knowledge, aspirations and skills of local residents (Evans and Foord 2006a). The rich mix in this project has largely been reduced to a commodified landscape of street retail and entertainment – a consumption opportunity for adjoining office workers, weekenders and the new urban professional. Although many of the ethnic quarters of most European cities now have Afro-Caribbean and Asian cultural centres, Jewish museums and multi-cultural arts centres, these spaces are 'predominantly independent and alternative to mainstream white-European cultural institutions. Few of these ethnic cultural spaces are flagships in the manner of the established museums and theatres of (high) European art' (ibid., 70).

Creative Quarters and Post-Industrial Regeneration

Although less visible than heritage and ethnic quarters, the rediscovered city quarter is also being drawn into the place branding mix in the form of the creative industries production district. Whilst cultural production has featured within heritage quarters, for example, Rope Quarter, Liverpool; Cultural Industries Quarter Sheffield; Jewellery Quarter, Birmingham; Liberty Village and Distillery District, Toronto, due to the availability of initially cheaper and flexible industrial premises (large and small floor space, lesser noise controls, etc.), as well as consumption spaces (art and entertainment, cafes, clubs, street markets, etc.) – the value of creative production in new sectors such as the digital economy has extended the city brand landscape (Evans 2009b), as the following case studies illustrate.

Amsterdam-Noord

Amsterdam 'North' is an emerging creative quarter in the area north of Amsterdam, a former working-class district separated from the centre by the IJ river. Literally the back door of the city, with access behind the main rail terminus, free ferries operate 24/7 with short 5- to 15-minute journeys across the water. Here, both new-build and re-use of former industrial buildings have facilitated a creative zone combining workspaces for ICT and media firms, ranging from the A-Lab managed workspace, to *Tolhuistuin*, a unique multi-use arts and entertainment venue comprising a concert hall, theatre and gallery spaces; whilst nearby, several floors of the former Royal Dutch Shell building – a 100-m-high, 22-storey landmark – has been converted for dance clubs and all-night events (Figure 5.3). The striking Eye Film Institute/Cinema building was also opened in 2012, relocated from its former museum quarter site on the mainland. Along the waterfront, also connected by free ferry, is MTV's Benelux HQ, as well as artists' workspaces in newly converted warehouses, including a 20,000 m hangar hosting an *Arts City* of makeshift studios. This previously nondescript district now hosts thriving cultural hubs that have emerged from long-term urban regeneration projects, the largest being the NDSM Wharf. Once an immense shipyard, this area is now one of Amsterdam's most popular off-beat locations, home to artist's studios and creative-digital businesses, whilst still retaining its industrial, factory-like aesthetic.

Figure 5.3 Amsterdam-Noord (Tolhuistin Pavilion; Eye Building and Former Royal Dutch Shell Tower)

Source: Photographs by the author.

This regenerated quarter of the city combines cultural activity with creative industry production and entertainment, and it will be interesting to see how these will coalesce into a zone for work and play over the new few years. What this combination of entrepreneurial, institutional and post-industrial placemaking does demonstrate is that new quarters of a city can still emerge in areas otherwise over-looked, effectively extending the city whilst relieving pressure on over-crowded and commodified cultural and tourist districts. The relocation of cultural flagships to more diverse areas can also present an interesting strategy for the cultural plan-ning of the city and present a radical alternative to the increasingly sterile museum and cultural quarters that have resulted in over-concentrated, mono-cultural areas of many cities. A sign of the competitive nature of these digital clusters, however, is the relocation of Google's European headquarters from Amsterdam to London, attracted by the Tech City brand (see Chapter 6). This underscores the inter-city competition that also saw the transnational Royal Dutch Shell company choose between the Netherlands and the UK for the location of its HQ (for tax purposes). In choosing London over Amsterdam, the company thus forfeited its historic Royal Dutch brand but left a structure that soon found usage as a cultural space.

Canada

Examples of creative industries quarter development in Canada include the *Cité Multimedia* in Montreal's redeveloping industrial waterfront district and the Lib-erty Hall complex in Toronto. In the King-Liberty area, a liberal approach to change of use was combined with restoration of building facades. Following the closure of former factories and warehouses, the area provided a natural incubator for small enterprises initiated by artists and designers. Low-rent premises were adapted for studios and workshops, including live-work accommodation, sometimes in contra-vention of planning controls. By the early 1990s, there was a significant policy shift away from the presumption that industry and housing were incompatible in close proximity. Rigid zoning for industrial use in the City Plan proved a structural con-straint for the emerging strategies for regeneration. As in the USA and Europe, the recycling of brown-field sites and deliberate creation of mixed land-use neighbour-hoods on the fringe of downtown, especially those incorporating cultural indus-tries, came to be seen as desirable aims of planning intervention. Flexible leases combined with easy accommodation of physical expansion enabled some to pros-per within an artistic community – both resident and mobile. New media industries moved into this area, serviced by bistro-style bars and restaurants, and employment increased by over 10%.

In francophone Montreal, a different approach to the development of creative industries quarters has been pursued. The city-region had been a prime manufac-turer in textiles and related production, but as this declined, Montreal had not devel-oped a specific design capability (unlike say Northern Italy and Scandinavia) which could switch to other forms of creative industry, for example, new media, designer-making fashion and textiles. The refurbishment of former industrial premises in downtown/heritage districts and waterside areas is therefore being supplemented with new-build premises to house multimedia firms in order to capture this grow-ing activity, as many other post-industrial cities have done. Grants for firms linked to employment encourage growth over a ten-year period, but the rentals and lease/purchase costs are at commercial rates. Labour and skills are being supported, rather than premises which have been the model elsewhere (e.g. managed work-spaces, subsidised rents/flexible lease terms). The logic in Montreal is that if the firm is successful in developing a service or product, which will be competitively financed due to subsidised labour costs, income/profits will be sufficient to pay higher rents and over time increase to attain self-sufficiency once employment sub-sidies ends. *Cité Multimedia*, on the other hand, was a Quebec government-led regeneration venture targeted at new technology firms, which were rolled out in eight development phases. Tax incentives attracted new and established hi-tech firms, but after the bursting of the dot.com bubble and the elimination of tax incen-tives for information technology jobs in the district in 2003, further phases were cancelled, and the remaining structures were sold by the joint venture to the private sector. Today over 6,000 workers are based in this area which attracts high rents. Interestingly following this policy shift, the Quebec government launched an IT cluster programme for medium-sized cities and rural areas of this Province, rather than centralising ICT R&D in the (cultural) capital of Montreal, and as Duxbury observed:

> [T]here is a disturbing absence of culture in the new visions for Montreal as a 'City of Innovation' and 'Knowledge City' . . . it appears that cultural and heritage activities and resources are recognised and valued insofar as they attract the scientists, and other knowledge workers the city is recruit-ing. However, cultural activities are not seen as part of the knowledge and innovation milieu itself.
>
> (2004, 1)

This raises the question, the extent to which the digital industries here can be con-sidered cultural or even creative industries – or indeed cultural spaces. Rather, this model resembles a closed, exclusive technological space, and applying the moniker 'creative' to this 'class' is therefore a misnomer.

The Distillery Historic District, Toronto

The potential symbolic and economic value of former industrial heritage buildings is now well recognised (Zukin 1995). They can provide attractive and interesting spaces to accommodate creative and cultural activity – both exhibition/entertain-ment based and production based. As Jane Jacobs argued: 'old ideas can sometimes use new buildings. New ideas must use old buildings' (1961, 188). Jacobs herself had emigrated from the USA to Toronto in the late-1960s, and her seminal *Death and Life of Great American Cities* (1961) formed the basis of her continued inter-est in urban design, mixed-use and the importance of a vibrant urban environment. The Distillery District is strategically located linking the city's downtown and waterfront areas with a mixed use/tenure neighbourhood, as promoted by Jacobs (1961). Once home to the Gooderham and Worts Distillery which finally closed in 1990 and since then a designated national historic site, the complex was redeveloped in 2001 as a pedestrian-only village almost entirely dedicated to arts, culture and entertainment. Work was completed and the district reopened to the public by 2003 and the new owners refused to lease any of the retail and restaurant space to chains or franchises, and accordingly, the majority of the buildings are occupied with independent boutiques, art galleries, restaurants, jewellery stores, cafés and coffee-houses, including a well-known microbrewery. The upper floors of a number of buildings have been leased to artists as studio spaces and to office tenants with a creative focus (Figure 5.4). The Distillery District also houses one of the city's largest affordable workspace developments for artists and arts organisations operated by the not-for-profit Artscape organisation (Evans 2001). After a C$3 million renovation between 2001 and 2003, 60 tenants moved into the Case Goods Warehouse and Cannery Building. These include artist and designer-maker studios, non-profit, theatre, dance and arts organisations (Gertler et al. 2006).

An innovative partnership in the Distillery District is the resident Soulpepper Theatre and a collaboration with the Theatre School at George Brown College. A new facility to house these two organisations, the Young Centre for Performing Arts, was opened to the public in 2006. This 44,000-foot performing arts, training and

Figure 5.4 Distillery District, Toronto

Source: Photograph by the author.

youth outreach centre allows students and professionals to work, learn and live side by side, sharing work studios, rehearsal halls, wardrobe and scenic facilities. The Centre features eight performance spaces suitable for audiences of 40–500. The project's founders were aware of the shortage of theatre space in the city and therefore encourage other performing arts organisations to book the facilities for their work. The Distillery District not only provides space for arts development in the city but it is also now one of Toronto's top visitor destinations (Gertler et al. 2006). The district is also an important resource to the film industry. In the first ten years, over 1,000 films, television shows and music videos have been filmed on location in the district. Since re-opening, the complex has undergone substantial capital investment in upgrading and new facilities, but at the same time key anchor tenants (galleries, restaurants and retail) have vacated the site. Maintaining a diverse mix and affordability for a range of tenants has proved problematic. Meanwhile, the neighbourhood has attracted new investment and dwellers with new apartment blocks and *condos* adjoining the heritage site, including athletes accommodation developed for the Pan American Games held in Toronto in 2015. When completed, the local population will exceed 2,500 which will be increased with further residential development in the surrounding neighbourhood. Whether this mixed neighbourhood can retain its quality and distinctive brand will remain to be seen – the next phase may see this heritage visitor district evolve to a more local destination. This also rests on its connectivity to the downtown area of the city (Matthews 2010), but as Kohn suggests:

> [T]he image promoted by the Distillery District could be described as the commodification of decommodification . . . the Distillery District is best understood not only as a strategy for maximizing the returns on investment capital but also as a cultural response to globalization and de-industrialization.
>
> (2010, 359)

Liberty Village, Toronto

Liberty Village is a 38-hectare, inner-city mixed-use site of commercial, light industrial and residential uses (Figure 5.5). The area was traditionally a conglomeration of factories, prisons and ammunition storage in the early industrial era, and until 1858, it was the site of Toronto's Industrial Exhibition. The developer in partnership with the City of Toronto had explicitly branded the area *Liberty Village* and like in the Distillery District, had engaged the Artscape artists' studio operator to create managed workspaces for arts and media firms. Today most of the Village's century-old buildings have been retained and converted into commercial spaces that house a collection of creative enterprises in digital, fashion and interior design, media, advertising, high technology, printing, food and drink industries. New apartments have been created in some of the large converted industrial buildings. By 2001, new-build condominium apartments in Liberty Village had become sought after. The lifestyle advantages of King-Liberty were being promoted strongly by real-estate agencies, with strap-lines aimed at passing longer-distance commuters such as: 'If you lived here, you'd be home by now!', with parallels to Denver, Colorado's lo-do area: 'Kiss the "Burbs G'Bye!"'

Figure 5.5 Liberty Village, Toronto
Source: Photograph by the author.

Due to the strong presence of technology-intensive firms in the area, Liberty Village is almost completely wireless. For example, the Liberty Market building, one of the latest redevelopment projects in the neighbourhood, developed 300,000 feet of commercial, retail and studio space including a completely wireless network (a local tech company based in Liberty Village, struck a deal with the City of Toronto to wire the area). The Liberty Village Business Improvement Association (BIA) has played a leading role in protecting and promoting this creativity-industry employment area. Officially designated in 2001, it was Canada's first non-traditional, non-retail BIA, with a campus-style mixed-use layout rather than the high-street retail strip typical of most BIAs. The LVBIA is funded by a special tax levy collected from commercial properties in the area. Businesses in Liberty Village automatically become one of the 500+ LVBIA members, representing more than 7,000 people who work in the district. The LVBIA also endeavours to improve and enhance the design, safety and security features of the area. It also acts as a liaison with the community through newsletters and special events and expresses the community's voice on various issues. Liberty Village's critical mass of creative entrepreneurs, wireless infrastructure and reasonable rents has been a major selling feature of the area to businesses looking for modern, innovative workspaces. The village consists of over 100 properties zoned commercial or industrial and is designated an employment zone in the City of Toronto Official Plan. The City's employment zones do not allow for residential use, with the exception of artists live/work studios. Nevertheless, the Village's popularity over recent years has brought on the pressures of residential, condominium encroachment, particularly to the west and north of the village. The Liberty Village BIA has worked with the area's City Councillor and the City of Toronto's Urban Development Services to review the planning of the area in order to provide direction for future development and identify where residential development would be most appropriate. This review also considered public realm, heritage, land use and transportation issues facing the area (Gertler et al. 2006).

However, as Wieditz observed: the area's makeover is supported by newspaper articles that promote the area as an 'artsy loft district', a 'bohemian enclave' and a 'neighbourhood to live, work and play' for people who want to be close to the entertainment district and to the gentrifying Queen Street West area. With the influx of large-scale developers, 'it is likely that the new developments will obliterate any trace of the 'artsy' and 'bohemian' residents who once populated the area' (2007, 6). Maintaining mixed use and a mixed economy and providing infrastructure (e.g. public transport) that supports the new dwellers to these regenerated areas, therefore, remain the prime challenges to these renewed urban quarters and in sustaining their distinctive brand. In Liberty Village, evidence of displacement has been apparent for several years – 'while artists and photographers were the first tenants to occupy abandoned industrial space in the area and were certainly seen as vanguards of the economic success of the village, they are rapidly being displaced by new media, television, advertising and design firms . . . artists were also evicted from 9 Hanna Avenue, a "mythic" building where artists, artisans, musicians and other activists lived and worked, often illegally. The building was redeveloped as a wired high-technology complex' (Catungal, Leslie and Hii 2009, 1108).

Hackney Wick and Fish Island

Gentrification and intense residential property development have also been a particular challenge for this east London creative industries quarter. Over 1,000 small enterprises are based in Hackney Wick and Fish Island (HWFI), with a third of these operating in sectors including music, performing and visual arts; film, TV, radio and photography; fashion and textiles design; advertising and publishing, employing over 650 people in salaried and freelance jobs (HWFI 2018). The area has been a traditional manufacturing district for several hundred years, including food (fish smoking, confectionery, fruit and vegetable growing), engineering and materials (plastics, electronics, ordnance) and an alternative cultural location for artists' studios, festivals and activism. Workspaces occupied today represent the legacy from this industrious and innovative past, including live-work and artist studio complexes. Under a Mayoral initiative in 2018, the area was designated a Creative Enterprise Zone (CEZ), one of several across London. The HWFI Creative Quarter is embodied by a consortium of cultural organisations and artists in the public, private and third sectors who have been key players in the emergence of the local artistic, cultural, creative and digital sectors. They have helped it gain an international reputation, but they recognise that the Quarter is at a pivotal stage. Key issues identified by the group include affordability of space at varying levels for artists, sole practitioners, independent traders, small and growing businesses; preservation of the character of the area; ensuring strong local leadership that is inclusive and representative; greater access for local people into cultural, creative and digital industry jobs; and supporting small enterprises to capitalise on the benefit of cluster supply chain opportunities.

The creative enterprises are self-organised through a Cultural Interest Group (CIG), which meets monthly throughout the year, an effective network that enables new entrants to connect and pitch their organisation to others, to lobby and campaign around local issues (local councillors can attend) and to commission research and promote the area's cultural economy. This has included an annual arts festival and other events such as open studios and tours. This governance model is more advanced than the more tacit and informal networks operating in industrial clusters and closed BIA/BID organisations (with membership by subscription/precept), but retains an informal style, but nonetheless a transparent and democratic system that is strongly associated with place and the history of this area. This is not in a sentimental heritage sense, but in terms of how this habitus has influenced, shaped and produced the area's present form and function, which is a large part of its appeal to cultural producers of all types and stages of development.

Mixed Use and Live Work

A particular feature of cultural production areas is the high levels of density they generate, which has enabled a more intensive use of space and creative interaction. This has especially suited smaller enterprises, flexible working, collaborative workspace arrangements and the support of a mixed-use/mixed economy system, reflective of cultural production and innovation. Cultural facilities, whether consumption or production based, represent the micro-level cultural space, whilst creative districts and regional systems can be considered the macro-level, including polycentric clusters that operate within the city, as well as city-regional and intra-regional scales, including transborder clusters (e.g. Euregion – van Heur et al. 2011). Meso-level space use, on the other hand, is represented by mixed-use buildings and a morphology (i.e. former industrial buildings) that lends itself to both vertical and horizontal integration (e.g. living above the shop, live-work studio, ateliers) and a concentration of space usage that enables temporal, social and economic mix. It is no accident therefore that the cultural and creative industry quarters described here contain a high proportion of mixed-use buildings, including live-work premises that lend themselves to studio living and working, for example, artists, designer-makers and co-location across particular creative production chains (e.g. *Creative-Digital*, as discussed in Chapter 6).

In the seminal Green Paper on the Urban Environment, the Commission of the European Communities first promoted the 'mixing of urban uses – of living, moving, working', taking as its model the old traditional life of the European City (e.g. Vienna, Barcelona, Berlin, Budapest) stressing density, multiple use, and social and cultural diversity (Commission of the European Communities, 1990, 43). This has been a concerted move towards the idea of the compact city (Foord 2010; Breheny 1996), with higher levels of land-use and residential density, and a move away from the zonal spatial separation of social and economic activities. Mixed-use development and a partial return to city centre living have been two consequences of this change in planning policy, property markets and lifestyle choice, particularly for early retirees and younger professionals (Evans, Aiesha and Foord 2009). This has also provided an opportunity for more sustainable urban economies with reduced travel/commuting and a range of services and amenities in proximity to work and home (Table 5.2).

Cultural production districts and managed workspaces – including artist studio complexes and live-work premises – have however operated in light industrial and, in some cases, semi-derelict areas of cities, prior to the zonal separation of land and economic use that had defined urban growth throughout the industrial era. So their re-discovery has generated the attention they now receive through city regeneration, place branding and re-use opportunities (workplace, residential, retail and leisure uses). The intensity of employment activity and the networks they support

Table 5.2 Advantages of Mixed Use

Concentration and Diversity of Activities		
Vitality	Less need to travel	Local Economy and Clusters
A more secure environment	Less reliance on car	Production chain; Innovation spillovers
More attractive and better quality town centres	More use of and opportunity for public transport, cycling, walking	More local employment, amenities and services
Economic, social and environmental benefits		

have also attracted social and cultural uses, such as bars/cafes/restaurants, music clubs, galleries and specialty retail shops, which in turn has made them attractive to visitors. Temporal uses, facilitated by more relaxed planning and licencing rules (and enforcement), have also meant that spaces are used for different activities at different times of the day and days of the week – so a daytime shop may also operate as a gallery or a music club at night or on the weekend, whilst a club may be used as a workspace or conference venue during the day. Night-time and late-night activity also flourished for these reasons, further extending economic, social and cultural activity in a way that formal cultural venues and town centres could not. This perhaps reached its zenith in cities such as post-unification Berlin (Fuller et al. 2018) and city fringe areas of London (Evans 2014b), where this combination of contributory factors – endogenous and exogenous – conspired to create the eponymous 24/7 imaginary, at least for a while. Whilst Berlin had benefited from the vacancies/voids, dereliction, freedoms and latent demand for work, entertainment and cheap living spaces in areas such as Friedrichshain-Kreuzberg, London's historic Clerkenwell district had retained its cultural and crafts production facilities in proximity to the city and new city residents of apartment and loft style living, and a unique presence of covered markets that operated throughout the night. As illustrated in Figure 5.6, Clerkenwell developed a particular vertical integration of building uses as well as a local-area mix of residential, workspace for small crafts, cultural and creative industries, and a range of amenities, entertainment and institutional facilities (university, hospital and market) that supported and served a residential community all in very close proximity.

The vertical space-use analysis is indicated by floor, and in terms of land-use types, based on analysis of building types/morphology, road/pedestrian networks and land-use mix. This shows not only the north-south separation between residential and more commercial/mixed-use zones but also the residential use above first floor, as the darker shaded blocks expand upwards with high-rise dwellings and encroach into the more mixed land-use area with a range of retail, offices, showrooms (e.g. furniture, interior design), cafes and other amenity uses on the ground-floor level (Figure 5.6).

With this mix and a growing daytime, evening and residential population, all-night clubs opened next to the large Smithfield (meat) market where night-workers clocked off and traditionally (i.e. for over a century) were then able to drink in local pubs from the early hours as clubbers were leaving their venues, and if they had the stamina and pockets, join them. These ostensible cultural quarters thus evolved a kind of double life, but inevitably once the balance between work and play was undone by excess and over-use, rising crime/anti-social behaviour and a disconnect between residents, workers and visitors (and ironically a more mono-use of these spaces), the night-time/24-hour economy boom in many cases has had to be curtailed, as the more liberal controls were reigned-in. However, by this stage, the property-led gentrification and placemaking by city and commercial operators had conspired to reduce both the affordability and appeal to the benign cultural producers that had organically, and in an un-planned way, created these mixed-use areas in the first place.

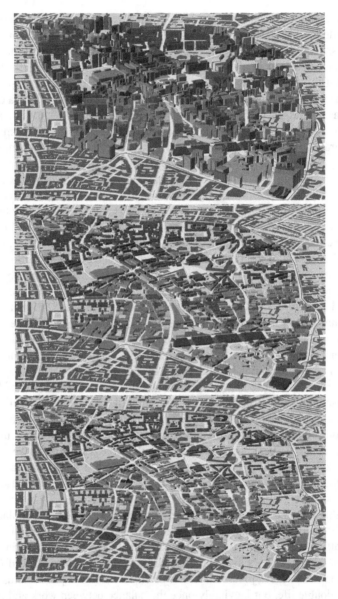

Figure 5.6 Clerkenwell Land-Use: Ground, First and Second Floors and Above

Placemaking or Making Places

Whilst creative hubs associated with former cultural and heritage quarters indicate the importance of historical and symbolic association and building types, which can transform and recreate new creative districts, it can be seen that zones can also emerge from post-industrial areas of cities that otherwise lacked placemaking potential. These former industrial city workshop areas, such as in Clerkenwell London; SoHo New York; and Eagle's Court, Adlershof Berlin (now a more university-science park model), thus benefit from proximity to the city centre in city fringe locations that effectively extend the city's visitor and cultural footprint. For example, the designation of the Poblenou district of Barcelona's new university-media-hi-tech quarter *@22* as an industrial (as opposed to services/education/housing) zone reflects its transition from a textile-production district to a *re*-newed creative industrial quarter (Charnock and Ribera-Fumaz 2014), but one largely lacking the symbolic, crafts and cultural economy (in fact established cultural firms are replaced by higher-economic value tech/professional services' use). What is clear in these scenarios is that the creative, media and digital industries require industrial planning and zoning protection no less than their industrial manufacturing forerunners and that an industrial brand is no less effective than a heritage or primarily consumption-based place. As can be seen, this *quarterisation* can in turn widen further, as adjoining districts join the spread of regeneration and cultural redevelopment, as witnessed for example in Brooklyn, New York, with the nascent creative-turned-consumption quarters of Greenpoint, Williamsburg, Red Hook and Dumbo – and in London's inexorable spread east towards the post-Olympics zone at Stratford and beyond ('Thames Gateway'). Connectivity is therefore important in creating and sustaining new quarters, as is a certain degree of distinction between these areas. This can be reflected in terms of legacy and historic association (e.g. place of origin branding); physically through the morphology and architectural quality and style (including maintaining mixed-use at block, street and area levels); through ethnic or other cultural experiences, for example, festivals and food; as well as in terms of particular cultural activities and business, for example, speciality retail, street markets and open trade fairs.

Placemaking through physical interventions and re-imaging the cityscape is evident in both incremental and radical forms, although this is generally less an act of deliberate branding, more a consequence of new-industrial development. Larger and established cities use this strategy to enhance and extend their cultural offer, and to signal their continued growth and confidence in the future. Spatially this is also used to create images and welcoming spaces for under-developed and 'edgy' zones and sites (Hannigan 1998) as part of placemaking and festivalisation efforts (Palmer and Richards 2010). At the other extreme, new or resurgent cities look to major works and spectacular architecture to impose a new physical brand and city image, often a high-risk strategy where little or no vernacular, pre-existing or cultural content is available. Both approaches tend to be top down in their execution, however, accompanied by what are token consultation processes (Evans 2005, 2006). Fifty years before, Lefebvre had lamented that participation is often invoked in city politics but rarely practiced seriously. In his time, as in ours, participation was impoverished. Citizens rarely had more than a nominal and advisory voice in decisions (Purcell 2013, 150) and Lefebvre called instead for '"real and active participation," (and) the pervasive activation and mobilization of inhabitants' (1996, 145), although not articulating how this might be realised. In these cultural production districts, networks and communities of interest within these spaces have provided some voice and a degree of autonomy over development, but over time this has proven difficult to maintain or to provide a force with enough clout against the liberal government regimes and rentier operators.

The cultural and creative quarter, on the other hand, presents an alternative dimension to more instrumental placemaking. This is not only due to their organic and largely unplanned development trajectory but also as a result of a combination of a symbolic/historic production culture and vernacular spaces, and entrepreneurial, even pioneering spirit, led by individuals or often not-for-profit agencies such as workspace providers, arts and heritage organisations, as well as small enterprise networks. In some cases, this includes Business Improvement Districts/Areas (BID/BIA) based around cultural and creative industries and festivals, for example, London's South Bank and Oscar Wilde's *Reading Gaol* (Historic England 2016). This more bottom-up revitalisation of city space has coincided with the growing pursuit of greater authenticity and the co-creation of everyday experiences by residents and visitors to these new and rediscovered quarters of the city (Richards 2019). As the examples provided here demonstrate, however, they share a common experience of commodification by the very bodies that exist to protect and maintain their productive status, a skewed type of creative destruction whose consequences may be unintended. However, in light of the accumulating evidence surrounding this process of place branding cultural production spaces, the impact of this treatment can no longer be in doubt, questioning the intentions of those pursuing place branding of cultural and creative districts. When these fragile and often historically based ecosystems decline through crowding out (economic, social and property), they cannot be re-invented by institutional and corporate versions (e.g. business and science parks), since these are never able to recreate the cultural foundations or organisation of the originators. Creative firms, of course, also exploit this branding process whilst they can, particularly where they seek to present a positive image towards their (potential) clients (Hutton 2006): 'for all entrepreneurs the physical environment is important in reproducing and strengthening their and their companies reputation of being creative – In this sense, place becomes a marketing device' (Hebbels and van Aalst 2010, 359).

Cities are complex, messy and with cultural forces that are lived and experienced from below, at street level, often colliding with notions of brand in the product and corporate sense. This can be seen as a positive attribute, since the enduring identity of a place is one that is shared but differentiated in the minds and experience of different communities of interest. Their perspective, literally and psychologically, of their city – whether they currently live or work there or not – makes up the collective identity of a place, which can also encompass resistance and obduracy (Hommels 2005) and also the adoption over time, of changes to the urban landscape. *Quarterisation* through cultural placemaking (Jayne and Bell 2004) as an aspiration rather than a *dirigiste* act should therefore form part of, but be subsidiary to, a cultural planning approach to the city, which is both sustainable (Evans 2013a) and a product of cultural governance and comprehensive mapping and use-valuation of a city's cultural assets (Evans 2008). Supporting, rather than exploiting cultural and creative quarters, would seem to have an important role in giving substance to this process and, where appropriate, placeshaping efforts. As Mommaas maintains:

> city branding not only meets the increased need to make one's own city stand out in the midst of an expanded and more mobile reality, but at the same time it also meets the need for sources of urban orientation and identification . . . lead(ing) to the creation of a need to refill a fragmented space with positive meanings that can function as new sources of civic pride.
>
> (2002, 44)

The question here is one of power over place and who holds this power and its distribution and what process is available to the civitas in the shaping and image projected of their place identity. This in turn implies practical not just conceptual rights and consideration of the role of culture and spaces for innovative thought in this creative dialectical exchange.

*

Digital-oriented production and spaces increasingly drive and feature in old cultural quarters and mixed-use areas of the city, leading to the emulation of the Silicon Valley imaginary by place branders and investors (government and commercial), but like cultural spaces generally, their specificity and particularities have generated quite distinct examples of innovation through local networks and placemaking. Digital cultural space is therefore the subject of the next chapter, as cultural space adapts, and is transformed through both physical and virtual realities.

Note

1 http://www1.geo.ntnu.edu.tw/~moise/Data/Books/Social/08%20part%20of%20theory/
henri%20lefebvre's%20the%20production%20of%20space.doc

6 Digital Cultural Spaces

The advent of digital ubiquity – the so-called *internet of things* – has in many respects presaged a decline in real-life/real-time cultural space and experience. Certainly, key aspects of cultural consumption and production have moved significantly online – notably in retail domains, including the symbolically and practically important music equipment and record shops, and also in traditional printing and publishing, whilst new digital technology such as computer-aided design, laser cutting and digital printing, has transformed the visualisation and manufacture (CAD/CAM) of fashion goods, jewellery, architecture and product design, among others. New media forms such as video games, already made obsolete by online and streamed gaming, have expanded screen-based activity and also its location, whilst collections in archives, museums and galleries are increasingly digitised for reference and to enhance and supplement the visitor experience:

> over the past 30 years, museums, galleries from the Metropolitan Museum in New York to Museum of Islamic Art in Qatar, from the National Museum in New Delhi to the tiny Lynn Museum in Norfolk, have put much of their collection online, making them available to millions, cultural treasure that would otherwise be denied them.
>
> (Malik 2022, 48)

A sign of things to come in the physical versus digital debate is also apparent with the Museum of Modern Art (MoMa) in New York deciding to auction off 29 of its masterpiece paintings – including by Picasso, Monet and Bacon – to help establish an endowment for digital media and technology that could be seen as the start of a virtual museum and the decline of the real cultural space and experience itself.

The extent to which digital cultural spaces differ fundamentally due to the technology interfaces and distance that online communication demands and offers, first, needs to consider production separate from consumption (Wolff 1981). What we can observe is a continuum and incremental change since reproduction of images and sound was enabled through photography, then film and broadcast media to today's online media. Consumption likewise has followed this trajectory with varying levels of agency, collective engagement and individual access (radio,

DOI: 10.4324/9781003216537-6

TV, camera, computer, etc.). A distinction, originating with Adorno (1991), is made between the traditionally pre-industrial creative processes, which then employ mass reproduction and distribution methods (e.g. books and records) and those where the cultural form is itself industrial (e.g. newspapers, film, television). This cultural industry definition is largely separate from the traditional performing and visual arts and therefore from the notion of arts amenity and cultural places which Garnham (1984) does however include in the core activities that contribute to his *transmission of meaning*. So that whilst the cultural content maybe familiar and, as we will discuss later, still retains key production characteristics, the consumption and opportunities for exchange (social, cultural, commercial) have fragmented and expanded in form and function *vis* mobiles, streaming and social media. The debate over the implications of the 'mass overproduction unleashed by mechanical technology' (Pickering 1999, 181) and its amplification by digital technology centres on issues of authenticity that Adorno and Horkheimer decried in the threat that Hollywood made as a dominant form of cultural [industry] globalisation (1943). Walter Benjamin, of course, had earlier addressed this detachment of art from traditional aesthetic and cultural values through the separability and transportability of the photographic image (1979). On the other hand, Miège (1989) and others challenged the critique of technology-driven mass culture. Instead, in the context of post-1968 social movements, he argued for recognition of differences across the many sub-sectors of the cultural industries, preferring the plural term 'cultural industries', recognising 'complexity, contestation and ambivalence' in the study of culture and maintaining that there were roles for culture beyond capital (Hesmondhalgh 2012, 45).

The evolution of both hardware and software applications in the cultural sphere has also had a much longer gestation than is realised, and this has exhibited distinct place-based phenomena which have more in common with other cultural space development – from Silicon Valley in California to Silicon Roundabout, London. Cultural production in the digital era also retains relationships, structures and place-based preferences which show a continuity with past cultural space location and usage – ateliers/studios, 'old' cultural and entertainment districts, university R&D facilities and workshops. This chapter therefore explores the extent to which the digital *Information City* (Cheshire and Uberti 2016; Downey and McGuigan 1999) is represented in terms of cultural spaces and the partial transformation of creative production that has arisen – albeit unevenly. In real and as well as in virtual cyberspace (the space behind the screen), 'it is possible to visit, look, listen, communicate, meet, buy, browse and simply hang out . . . the citizen can be participant, explorer, worker, consumer or merely flaneur' (Pickering 1999, 181).

The history of digital cultural spaces has yet to be written. Some digital-media-inspired retail experiences have had a short shelf-life, particularly where emulating the theme-park with the risk of theme-decay and regular, expensive updating – the *new ride*. Digital production in the live arts has swiftly moved on from blue-screen backdrops, enabling real-time collaboration across space/place as in music and dance, to immersive theatre, interestingly low-tech, for example Punchdrunk

Theatre, who perform in non-theatre venues and locations such as disused factories, post offices, quaysides, to audiences who roam through these 'non-arts' spaces:

> Amazon taught us we don't need to go to the shops, the pandemic has taught us we don't need to go to the office, Punchdrunk taught us we don't need to sit still in a theatre, and now they are proving that you can be in the play without going to the theatre at all.
>
> (Alex Mahon, CEO of Channel 4)[1]

The assumption that immersive experience has to be digitally delivered is clearly not necessarily the case. On the other hand, the Royal Shakespeare Company was an early adopter of Virtual Reality (VR), staging a live online performance using VR technology. The performance, titled *Dream* and based on Shakespeare's *A Midsummer Night's Dream*, was set in a virtual forest, with motion sensors on the actors allowing them to interact with their surroundings and the audience at home. Through their tablet, mobile or the show's website, audiences directly influence the live performance from wherever they are in the world. Digital technology has also transformed stage set design and lighting (still within traditional performance venues), with dance, more so than voice, benefiting from hybrid dancer/holograph/avatar combinations, where the dancer is 'part robot, part house diva, with sheets of light slicing the stage and complex digital projections suggesting neural networks and electronic circuitry chasing the dancer across the floor' (Winship 2023, 19). More holographic use of VR technology also allows artists to be youthfully renewed and resurrected, for example ABBA's holographic *Voyage* production, and in spectacular shows such as the opening of the Beijing Summer and Winter Olympics (2008/2022) which used 5G, Artificial Intelligence, AR/VR and cloud technology to create an immersive 3D experience for the spectators, although the use of AR/VR had in fact been developed and delivered by advertising and digital media firms, for example, in fashion shows[2] combining real and avatar models and visual display boards, several years before being explored on the public stage.

An earlier foray into cybernetics, in a cultural centre context, had featured in the architect Cedric Price's 1964 concept for theatre impresario Joan Littlewood's Fun Palace. This was planned to be located on an island site at Mill Meads near Littlewood's Stratford Theatre (now the site of the Aquatic Centre in London's Olympic Park) based on a design model that was prescient in many ways: 'temporary and flexible, with no permanent structures, no concrete stadia stained and cracking, no legacy of noble architecture, quickly dating' (Littlewood 1964, 432). Price's vision was for a 'new kind of active and dynamic architecture which would permit multiple uses and which would constantly adapt to change, thinking of the Fun Palace in terms of process, as events in time rather than objects in space' (Mathews 2005, 40). The building would have no single entry point and divide into activity zones (Evans 2015b). Price and Littlewood had assembled a multi-disciplinary team from architecture, art, theatre, technology and even situationists, with cybernetics and game theory driving the facility's day-to-day behaviour and performative strategies which would be stimulated through feedback from users. This marshy site would have been expensive to reclaim – although public funding was, of course, found subsequently for the bottomless finances accessed for the Olympics and ongoing legacy. The Fun Palace idea was also the victim of London's reorganisation into 33 boroughs with the London County Council transferring the open spaces to a new benign Lea Valley Park Authority, with a different perspective on fun – and design (Evans and Reay 1996). Although never built, Price's design concepts were widely influential, notably in the Pompidou Centre, Paris, and a smaller fun factory, the Inter-Action Arts Centre in north London (Chapter 2), designed using modular cabins hung from a steel frame which could expand and contract according to need. Despite the hype, however, the attempts to engage users through programming, architectural design and immersive experiences contrast perhaps with the limited offer provided by VR/AR, such as the *Metaverse* and other proprietorial digital platforms – hindered by clunky physical hardware and, critically, a lack of cultural content and authenticity.

Time-based media and other electronic and digital installations have, of course, populated museums and galleries since the 1960s with artists' film a new genre also emerging from this time (Curtis 2020). The autobiographical turn in visual arts has also lessened the symbolic and literal association with place, including the celebration of the domestic and individual/celebrity as subject (e.g. Gilbert and George), and in feminist art, for example the works of Tracy Emin, Sarah Lucas and Cindy Sherman. Artists in their dotage have also returned to their 'roots', marking the inspiration or birth of their practice with hoped-for posterity through local galleries and foundations in their names (Emin – Margate, Gilbert and George – Spitalfields), whilst artists legacies are still celebrated through their work-homes (Holliss 2015), such as Frida Khalo's studio/house and garden in Mexico city; Henry Moore's house and sculpture park in Hertfordshire, Barbara Hepworth's house and garden in Cornwall; David Hockney's gallery outside Bradford; and Eduardo Chillida's sculpture park and museum in his Basque country homeland (less so perhaps, Steve Job's garage in Paolo Aalto). These authentic places of production are likely to endure more so than the virtual digital tours of their works. Tradition in landscape art persists, however, whether depicting urban or rural scenes, for example, David Hockney's pictures of his home landscape in Yorkshire, but these mid-1980s artists have also embraced the iPad and iPhone as painting tools and his latest immersive experience opened in London in a new four-storey space *The Lightroom*[3] that 'aspires to be visually astonishing, alive with sound and rich with new perspectives' (Khomami 2022, 9). Using AR and VR, visitors hear his voice talking to them through his work, old and new, from his (artist's) perspective. But as critic Jones observes: 'the hour long immersion in gigantic projections makes less impact than a brief glance at an actual original work of art by Hockney in a gallery ... it's a sad fact that in this kind of spectacle, photography and film clips have more reality than drawings and paintings, so Hockney in his innocence has lent his fame here to a dumb contemporary fad that doesn't – and cannot – capture the beauty of his art' (2023). This follows less intimate immersive media exhibitions of the work of the late Freda Khalo, Dali, Klimt and Van Gogh. To give Hockney the last word however:

> Cinema is dying. You can see any Hollywood film on a big screen at home. But this is a new kind of theatre, a new kind of cinema. With this, you have to go out and see it. And people do like going out.
>
> (Jonze 2023, 19)

The point here is that a digitally enabled cultural experience can be both collectively consumed and located in dedicated cultural space, but not as a substitute for the authentic subject or object (a similar situation in animated and digitally enhanced museum exhibits). The test will be the extent to which these experiences endure or become tired, like a theme park ride.

The transformation of consumption and production of cultural goods and services has therefore been profound in some areas, but marginal in others. Certainly the growth of the so-called creative industries owes much to the opportunities generated by digital and ICT applications and processes, with distinct winners and losers in employment terms, a notable decline in publishing, printing, crafts, manufacturing (e.g. records, textiles) and design employment and firms. On the other hand, 'creative-digital' work (see the following) has flourished in video/computer games (referred to as *leisure software* in official creative industries statistics, and the main driver of creative economy growth – DCMS 2001) and digital content creation, now referred to as *storytelling in the digital sphere*.[4]

Big Tech

Some of the roots of the digital revolution do however exhibit notable place-based legacies. Boston's Route 128 is now associated with large tech (e.g. General Electric HQ) and entrepreneurs emerging from Harvard and MIT universities (Rissola et al. 2019; Saxenian 1994), but back in 1916, Massachusetts Institute of Technology (MIT) which was founded in 1861 witnessed a spectacular cultural event to mark the opening of the New Tech campus building, accompanied by dancers, actors, a chorus of 500 singers and orchestra, and attended by E. D. Roosevelt and Alexander Graham Bell (Jarzombek 2004). Its Californian cousin, Silicon Valley, the US motherland of most of the platforms which have dominated the social/digital media industry[5] – Apple, Adobe, Alphabet-*Google/YouTube*, Meta-*Facebook/Instagram/WhatsApp* – also evolved from the original hardware roots of companies such as Hewlett Packard, Rank Xerox and Sun Computers, in proximity to both R&D-based universities (e.g. Stanford) and military installations. It is no accident that their open innovation status was built on symbolic cultural spaces which economists would put down to a combination of comparative and competitive advantage, that is, skills/labour, clusters/agglomeration, innovation transfer and path dependency (Simmie et al. 2008). This is evident in the locations chosen for the headquarters of these global firms. For instance, Facebook recently moved its HQ from its Paolo Alto home to a site formerly occupied and owned by

Sun Microsystems in Menlo Park, California. Sun is owned by Oracle (Java), the world's second largest software company. This development entails new housing for Facebook workers and mirrors the Facebook 'village' it has created, but on a larger scale. Its Paolo Alto site was originally occupied by Hewlett Packard, a few blocks from Stanford University campus. This site will not be vacated when Facebook moves its HQ; instead, adjoining sites have been acquired for further housing development and amenities – coined *Zuckerville* after founder Mark Zuckerberg. Meanwhile, Elon Tusk, who relocated his Twitter HQ from California to Texas (citing excess bureaucracy), is building housing for his workers at *Snailbrook*, a 'Texas Utopia'.

The largest of the computer empires, Apple, has also undertaken a mega-HQ development – Apple Park, a 2.8 million square foot, $5-billion new-build circular building with a donut centre a mile wide, viewable from space. The previous occupant of the site for this starship, starchitectural extravagance was again Hewlett Packard, cementing the deep place-based culture of this evolving story (it was also Hewlett Packard that gave the Apple co-founder Steve Jobs his first job). However, with a large migrant population in the county (40% foreign-born), including first- and second-generation Hispanic (25% of the population) and Vietnamese, who 'service' this knowledge economy, Silicon Valley firms are not strong supporters of either the community or cultural programmes generally and remain in industrial parks on the city fringe. Most are new firms – exceptions being the long-established Hewlett Packard and the more recent Google with mid-West founders and a transient and different set of migrant workers (e.g. from south-east Asia). This 'plug-n-play' place, ranked highly and tellingly on both of Florida's 'Tolerance' *and* 'Inequality' indices (2005), is judged to be neither socially cohesive nor a vibrant (in fact a 'dull') place to live (Kriedler 2005). A further irony is the digital divide in access and ownership of ICT – in California, Latino young people were half as likely to have computer access at home: 36% compared with 77% of US-born non-Latinos (Fairlie et al. 2006). Like Sharon Zukin's narrative on New York *Loft Living* (1982), where artistic capital fed real-estate capital, Rebecca Solnit's visual documentation of the Bay Area's transformation through sky-rocketing commercial and residential rents driving out artists, activists and non-profits powerfully illustrated how the expansive imperial spaces of dot.com businesses led to a form of cultural impoverishment, the decay of public life, and the erasure of the sites of civic memory and disappearance of areas in which artists can live and create (Solnit 2001).

Silicon Somewheres and Everywheres

Given the perceived success of Silicon Valley as an innovation centre with rapid valorisation and capitalisation of its outputs, the pattern of technology districts adopting the prefix *Silicon* has not surprisingly accelerated over the last two decades. On the one hand, this is a case of both place and 'hard branding' (Evans 2003a) through the hope value associated with emulating Silicon Valley or *Silicon Somewheres* (Florida 2005). Unlike the microchip and hardware manufacturing sectors, however, these new media clusters are a digital-creative hybrid producing the cultural content that drives the numerous online platforms, web-based services and mobile devices.

Digital clusters that have evolved more organically can be seen at various scales – both regional and in highly concentrated spatial geographies. Examples include Silicon Fen (Cambridge, UK) and Silicon Glen (Dundee, Scotland), to local hubs where ICT firms often co-locate with creative and other advanced producer and financial services. Examples of the latter include Silicon Sentier in Paris; Silicon Allee in Berlin; Silicon Alley in New York – and Silicon Roundabout, or Digital Shoreditch, in east London (the roundabout having been removed in 2021 to be replaced by a pedestrian and bike-friendly square). Unlike the technopark or the impervious tech campus and industrial mono-estates that often litter edge city and occupy motorway-side sheds, these digital spaces are messy, heterogeneous and form part of a wider creative and cultural ecology of their vicinities. Indeed, they have often evolved from earlier cultural production practice and industry and operate initially unbranded and unmarked in mixed-use urban fringe areas.

Digital Shoreditch, London

This latter cultural-creative-digital district (Foord 2013; Cities Institute 2010; Evans 2019) presents a particularly novel and temporal digital cultural space, located in a city fringe area historically nondescript, with a low-income/deprived resident community, essentially a working area of the city previously untouched by the visitor economy or more conspicuous cultural consumption. Its cultural work-space tradition dates back several centuries to crafts – jewellery, metalwork, printing and publishing, and fashion and textile sweatshops – with an established artist community occupying affordable studio spaces, that is, a particular affordance that has supported and enabled place-based cultural production.

This low-cost cultural economy provided crucial elements in the area's transformation to one of the most celebrated creative quarters in the world. This now contains a high concentration of new media and digital firms, alternative nightclubs and venues for music, art and independent retail outlets and a high concentration of digital creatives or *digerati* (Foord 2013). A new destination has thus evolved in this east London district – several boutique hotels have opened in recent years including the ACE (the first outside of the USA, originating in Seattle, home of Microsoft), designed with materials produced locally from specialist bricks and tiles to lighting, and with photographic references in bedrooms to the building's music hall past.

A closer insight to this digital cultural space reveals an eclectic range of small firms whose founders came from diverse disciplines and interests, including art, design, architecture, music, publishing, as well as the computer-science-related specialisms more associated with tech firms. Most were what would be considered micro-firms, but this belied, on the one hand, their financial impact and success, and on the other hand, the network and project-based structure which vastly understated their scale of operation. Through the use of freelancers, sub-contracting and partnerships, locally, nationally and internationally, major contracts were undertaken by ostensibly small firms operating from shared and flexible premises. Valorisation of some firms was also evident through more traditional buy-outs by major media and tech companies and investors – for example, Last FM, a digital-media company which sold for £140 million to CBS, and Dopplr, a social-networking site sold to Nokia for over £10 million in 2009 (although by 2013 Dopplr had ceased to be operated by its new owner). Partnership and networking were the prime modus operandi, and this was facilitated through shared workspace arrangements in re-occupied former industrial buildings and start-up facilities run by local colleges and entrepreneurs. Monthly meet-ups became the key point of introduction and exchange for newcomers, offering the opportunity for pitches to be made, skills to be identified and exchanged and projects to be developed. This social opportunity also allowed larger established firms to meet smaller digital creatives and vice versa. The presence of bars, clubs and other informal meeting places in walking distance further enabled engagement and synergies to develop, including arts and cultural organisations and designer-makers longer established in the area, whose own production and communication methods increasingly required digital expertise and solutions.

This cluster, or more accurately network of small creative-digital firms and enterprises,[6] had been 'discovered' (*sic*) in the late-2000s first by the digital media press (*Wired*, *TechCrunch*), to be picked up by mainstream news, then local and national government, and as a consequence, international investors and corporate digital companies (e.g. Google, Microsoft, Vodafone, Cisco). However, this group of small enterprises had emerged in a swathe of city and east London's *City Fringe* from the 1970s onwards, arising from the so-called 'creative destruction' seen in the publishing, printing and graphic design sectors (Evans 1999a), although maps of firms across these two sectors still indicate significant presence and co-location between these sectors at this time (Figure 6.1). These distinctions between creative sectors also belie the fact that official classifications are rendered obsolete by innovation in production methods (NESTA 2012), since many print and publishing firms adopted desk-top publishing and digital graphic applications several years before the dot.com revolution. This was not entirely technology-generated, however, since restructuring within media, music, publishing, film and other creative sectors had also accelerated outsourcing/contracting-out to independent production, along with liberalisation of licencing which expanded the demand for creative content (Evans 1999b). This of course grew exponentially as digital-media platforms, and general web-based services further fuelled this demand for content. New enterprises and individual entrepreneurs joined this core group as the social media, games and web-based services grew, represented by a close network of over 200 primarily small firms. Intermediaries in the Digital Shoreditch network were low profile and not dominant and included a local university incubator, a further education (FE) college, and a small number of large companies historically located in the Shoreditch area, centred on Old Street (to be christened *Silicon Roundabout*[7]), such as Inmarsat, a long-established satellite company.

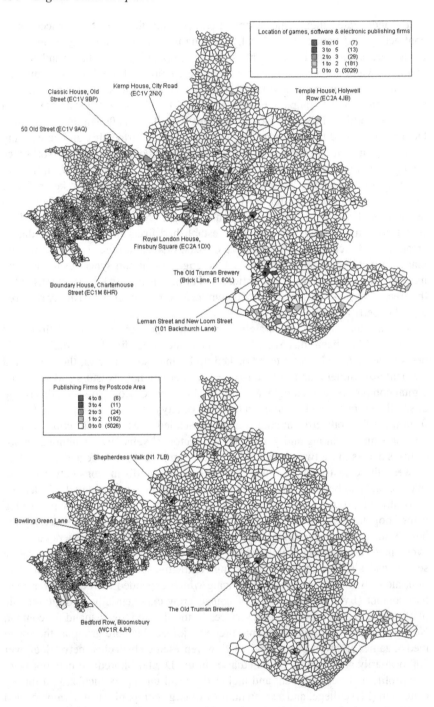

Figure 6.1 Print and Publishing and ICT Firms, City Fringe, London

Before any corporate attention to this nascent creative-digital hub, a local property entrepreneur and technology journalist had seen the opportunity of this new cluster through extensive publicity surrounding their founding of *Tech Hub*, a space that 'takes away the headache of having to sign leases' and where enterprising engineers can 'code with coffee in the company of likeminded people', describing this venture as motivated by a desire 'to put a flag in the ground, marking the territory of tech start ups' (Foord 2013, 54). This was a place-based, *placemaking* strategy (Evans 2014a) which chimed with a property-based approach to this cluster, providing another type of business culture and relationship with firms that needed both offices and informal spaces in which to inter-act (and away from predatory tech companies).

The point in terms of organisational experience is that the core founders of this digital network had an established entrepreneurial and independent culture, one that also drew on their experience of larger firms/sectors earlier in their careers, and as practising artists and designers, but also on lessons learnt in radical transformation, as technological, institutional and policy changes redrew the landscape of creative production and consumption. Successful in adapting and adopting new technologies and production chains, they had proved to be resilient and flexible and in many respects were 'new adopters'. Their large firm/self-employment/new business trajectory did however follow more traditional start-up routes. This contrasts with newer enterprises which typically lacked this large-firm experience, bypassing the corporate world, and reflected a younger age group – often post-university where existing personal network links were particularly strong and a market (business to business, employment, end-consumers) for digital creation was more established. This observation that these firms prefer *existing* networks rather than new cluster-facilitated relationships is important and is also seen in other creative clusters, such as in Berlin: 'rather than opportunities for informal networking, *already established* networks of friends and colleagues were vital in the entrepreneurs decision to locate in Prenzlauer Berg or Kreuzberg' (Hebbels and van Aalst 2010, 360), and these networks are not solely or primarily ends in themselves. They perform important functions to the aspiring and established entrepreneurs alike, albeit perhaps in different ways (ibid.) with more established creative tech firms relying less on contacts in their specific neighbourhood. The close proximity of this digital cluster to the City of London also enabled to them on the one hand to maintain their independence and alternative work and play culture, but when needed, to access the mainstream corporate world:

> London has the unique mix of tech, culture and banking. As the crow flies it is one mile from our base among the iPad toting hipsters of Old Street Roundabout to the pin-stripe suit wearers at Bank in the heart of the City of London. So Tech City really is uniquely located – it's hard to imagine a low-cost area full of talent only a stone's throw from Wall Street.
>
> (Andrew White, CEO FunApps, in Foord 2013, 55)

The emerging *fintech* sector nearby also offered a collegiate group and potential joint market with shared opportunities facilitated through the open Digital

Shoreditch network – but with a distinction visible in their attendance at local meetings (i.e. the *fintechs* wore suits and ties).

This profile and reputation also created a demand for hosting key design and digital events and festivals – from the London Design Festival to the week-long annual Digital Shoreditch Festival which was first held in 2011 attracting 2,000 participants/visitors rising to 15,000 in 2013. The demand for festivalising the digital economy is also another indication that digital production (and engagement) is neither placeless nor faceless. Tech strands at South-by-South West (SXSW) in Austin, USA (originally a film festival), and annual Tech EXPO (TECHSPO) in Amsterdam draw global audiences and 'in person' attendances in the digital age. The public face of the creative-digital production network was further extended by holding open studio events, emulating artist and architecture open festivals (Chapter 3), with digital studios and workspaces opening up to the public, holding coding and taster session for local school children. It was no coincidence that the first Open artist studios had been held in this area back in the 1970s (*Hidden Art of Hackney*, Foord 1999) and the first architecture biennale held in Clerkenwell in 2004 (Aiesha and Evans 2017), affirming the art, cultural and creative production tradition of this area, to which the latest digital incarnation is just the most recent addition.

The co-location of creative businesses is therefore seen to fuel co-operation, collaboration and, in particular circumstances, significant innovation (Boschm 2005; Bakhshi, McVittie and Simmie 2008). Where successful, as in the case of this digital cultural agglomeration, they draw in firms and labour working in similar and related sectors, increasing density through co-location and therefore maximising the benefits of a particular locality. Whilst physical proximity is one and most visible manifestation of a cluster, institutional proximity also describes how organisations are bound together through the same norms and incentives (e.g. shared political environment, cultural preferences). On the other hand, social proximity reflects the highly embedded personal relationships and connections between firms (Granovetter 1985), whilst cognitive proximity (Molina-Morales et al. 2015) represents the degree of mutual learning through a shared knowledge base, which is particularly beneficial to network dynamics. All of these proximity advantages are evident in Digital Shoreditch and in many other so-called cultural clusters (Evans 2009b). Although it would be more accurate to apply the concept 'network of aesthetic production' (van Heur 2010) rather than creative cluster, since this is the key *modus operandi* of these digital media producers, who co-locate in an area which is itself fluid and linked to polycentric clusters across the city and beyond (Cities Institute 2010; Evans 2009b). Another factor fuelling this digital cultural space is proximity to complementary congruent clusters (despite the opportunity of distance/remote working) – in terms of finance and legal expertise; advertising and marketing firms; universities in terms of R&D/funding and knowledge exchange/transfer activities; and to skilled labour, including students/start-ups – for example, art and design, coding, programming, visualisation, business studies and arts/event management.

However, according to Simmie:

> [T]he cluster idea has taken many academics and policy-makers by storm. It has become the accepted wisdom more quickly than any other major idea in the field in recent years at the expense of previous explanations and lacking in relevant empirical evidence.
>
> (2006, 184)

What these arguments and their evidence-base tend to omit therefore – drawn as they are from economic geography, regional and innovation studies – is the symbolic cultural, historic and heritage value that these cultural spaces also possess, which both attract and enable such places to flourish and remain distinctive. This includes architecture and the built environment, the mix of spaces and uses of these spaces (Evans 2014b), and a legacy of arts and cultural production and invention arising. In the much-hyped new digital economy, this includes early (and pre-web) technological development and innovation in fields such as publishing and printing, graphic design, music and recording, film and broadcasting, fashion and textiles, jewellery/metalwork and furniture/woodwork. These arts and crafts/ designer-makers had been operating in this city fringe area for several hundred years, then working outside of controls operating in the City (i.e. craft guilds, censors, licencing, planning) and benefiting from small workshop and shared workspace facilities including live-work or work-home production (Holliss 2015). The value of the historical legacy of these areas is not lost on new creative entrepreneurs, in the case of Berlin's Prenzlkauer Berg:

> The history which is still visible in the buildings, inspires me. Many things are still open, not finished yet, anything could happen and I can contribute to this. There are still so many unspoilt corners; you can live here just as you want; You can see the whole industry of life here, in separate layers. Even traces of a hundred years back are still visible. It isn't covered up, it's not polished.
>
> (Hebbels and van Aalst 2010, 358)

Breznitz and Noonan (2014) also study the effects of arts and cultural districts combined with the presence of research-intensive universities, arts colleges and the growth in the media arts sector. Using a range of descriptive statistics based on a sample of 89 cities with cultural districts, including employment change and patent data (shares and trends) over two decades, their analysis found that the presence of cultural districts and research intensive universities had little explanatory power across arts and media-arts employment. However, art colleges were positively associated with higher levels of employment growth in these sectors. Cities with cultural districts did see much faster rates in media arts patenting but surprisingly not the presence of universities, concluding that the more innovative cities in media arts appear to be those with an arts or cultural district (see *Natural Cultural Districts*, Chapter 9). This of course depends on the nature of cultural districts and other factors (as the authors admit), but it is interesting to note that in London, the biggest media-arts cluster, Digital Shoreditch, has evolved largely independent of either local university or arts school influence, with growth factors such as connectivity, personal networks (non-place-based) and proximity to production/supply chains in financial, advertising and other creative industries, in a pre-existing cultural district (Foord 2013; Evans 2019). This cultural hub had also been designated in both sub-regional city growth and city-wide creative-industry strategies (LDA 2003), recognising that cultural assets directly influenced the location and growth of creative and digital industries (TBR, Evans and NEF 2016).

The conditions, which may be historical ('heritage'), even spiritual or sacred – notions that still have lived relevance in non-Western societies/places – and which lead to a build-up of creative activity and invention, may also contribute to what Lee refers to in the spatial sphere, as a *habitus of location*. Lee, drawing on Bourdieu, suggests that cities have enduring cultural orientations which exist and function relatively independent of their current populations or of the numerous social processes at any particular time:

> in this sense we can describe a city as having a certain cultural character . . . which clearly transcends the popular representations of the populations of certain cities, or that manifestly expressed by a city's public and private institutions.
>
> (1997, 132)

This latter point is important in any consideration of economic development and cultural planning, since attempts by municipal and commercial agencies to create or manipulate a city's cultural character are likely to fail, produce pastiche or superficial culture, and even drive out any inherent creative spirit that might exist in the first place (Evans 2001).

Tech City

Nonetheless, the strategic importance attached to this sub-regional cluster and its role in the new digital industries was recognised in 2010 when the UK government designated the wider area anchored in the City Fringe by this creative industries quarter, as a classic urban imaginary, *Tech City* (McKinsey & Co 2011) – a swathe connecting this quarter further east to the Olympic Park, despite the absence of creative-digital firms in the eastern corridor, representing the physical legacy from the London 2012 Summer Games (Evans 2015b, 2016a). In April 2012, Google also opened its local *Campus* providing seven floors of co-working, flexible space, free high-speed internet, a Device Lab and Accelerator spaces, and of course, a Campus café, as well as daily events, regular speaker series and lectures, mentorship and training from local Google staff. Speaking at the opening of Campus in 2012, its head Eze Vidra said:

> We welcome members of the startup community: entrepreneurs, investors, developers, designers, lawyers, accountants, etc. and hope that this informal,

highly concentrated space will lead to chance meetings and interactions that will generate the ideas and partnerships that will drive new, innovative businesses.[8]

An additional layer was added to the earlier urban imaginary of the City Fringe, this time transforming its cultural production identity and edge location (Foord 2013). The City Fringe, once underpinned by craft and contemporary arts, was now synonymous with east London and tech start-ups and the central hub of a sub-regional Tech City corridor stretching further east. Campus London was Google's space for entrepreneurs with campuses now operating in Seoul, Warsaw, Tel Aviv, Sao Paolo and Madrid. Microsoft Accelerator's Campus quickly followed Google into the nascent east London Digital Shoreditch cluster of creative digital firms, to be followed by the Silicon Valley Investment Bank and public agencies, notably Tech City UK, formed by the UK government to capture and promote this urban Silicon Valley simulacrum.

Ten years on, Google's new UK headquarters are nearing completion less than three miles away in King's Cross, the international (Eurostar) railway hub (Partridge 2022). This 11-storey building will house 4,000 of the company's staff who will enjoy a swimming pool, a multi-games area for basketball, five-a-side football or tennis, a roof-top exercise trail and 'nap pods' for resting afterwards. This building will also include a community space for use by local residents and ground-floor retail units, some of which will be rented to small and emerging British brands. This vague 'community benefit' is common in large developments, in part to secure planning permission and placate opposition locally. Time will tell how these aspects will work when the building opens fully in 2024. Meanwhile, the Silicon Valley Bank that moved into Digital Shoreditch has been wound-up following the insolvency of its US parent, unable to service its cash flow in a febrile, high-interest era. The UK branch has been acquired by an international 'high street' bank, HSBC (formerly the Hong Kong Shanghai Banking Corporation), but no longer a venture capital lender specialising in higher risk small tech start-ups.

The transformation of old cultural districts to a visually digital industrial world can therefore be seen on the one hand as a natural evolution of old cultural forms to new ones as digital-inspired technology evolves and is adapted. What has changed however is the governance of these facilities, with large corporations, including media, IT, finance and property companies controlling these developments in key city sites, whether in Berlin or *Bengalaru*, Bangalore. What is clear is that they occupy the same historic and cultural spaces – part creative destruction and part globalisation, a seemingly irresistible combination.

Outernet – In(ternet)-Side-Out?

For example, *Outernet* is a brand (*sic*)-new facility recently built over the Tottenham Court Road station on the London Underground's new Elizabeth line. This area has been an arts and entertainment district for many years, bordering on London's Soho, a heartland for clubs (for dancing, drinking, music, stripping), music instrument stores and offices above for music agents, record companies, publishers and booksellers (art, antique, scholarly) – collectively known as *Tin Pan Alley* – as well as a music venue, the Astoria, representing a once-thriving hub for the music industry, from pop to punk (viz., Rolling Stones, Jimi Hendrix, David Bowie, Sex Pistols, Ed Sheeran, etc.). Walking distance from here, the first-ever internet café, Cyberia, opened in 1994, based on an idea originally conceived at the ICA arts centre. The ICA has been a radical space for new art movements and thought since the 1940s (Massey 2014) located incongruously on The Mall, in a neo-classical building leading to Buckingham Palace (see Arts Lab/Centres – Chapter 2). However, this coffee-house tradition would be hard to discern in the brash Times Square-style screens of this latest conflation of the old and new cultural space which describes itself as 'a radical new technology-driven marketing entertainment and information service housed in a super-flexible, digitally enabled streetscape' (Moore 2022, 28).

The 2,000-seat Astoria was demolished to make way for this complex. This had originally been built in the 1920s as a cinema, then converted to a theatre to be finally adapted as a live music venue (a familiar adaptation between the nineteenth and twenty-first centuries, from music hall to cinema, to theatre and music venue), becoming an iconic space which helped to launch the careers of many British rock bands over 50 years. Whilst the mix of uses has partially been retained (on the insistence of the local planning authority) notably music stores and a smaller 600-seat venue (@sohoplace) 'in spirit it is utterly changed . . . built on the obvious paradox that a culture fuelled by rebellion and chaos should now be channelled through the processes of large property owners' (ibid.)

Describing itself as an immersive entertainment district where music, film, art, gaming and retail experiences come to life in new breathtaking ways, it is in fact a single project in the ownership of one company, the aptly named Consolidated Developments. This is a far cry from the cultural clusters and Digital Shoreditch described earlier, which are resolutely not corporate, made up of many smaller and a few larger enterprises, workspace providers, cultural organisations and entertainment venues, and happy to occupy and adapt existing buildings, unlike global tech company villages and HQs. Outernet's digital credentials rest on an atrium with 23,000-square-feet floor-to-ceiling high-resolution LED screens, whilst other spaces surround visitors with more screens on which sponsored brands will promote their products, akin to a giant walk-in billboard (Wainwright 2022), revenue from which will help subsidise the few remaining music shops at the rear (Figure 6.2). In several respects, this new digital experience resembles the electronic billboards of Times Square, Piccadilly Circus and in the downtowns of Tokyo and other major cities, a neon version of the upmarket graffiti that Walter Benjamin had distinguished (1999 and see Chapter 8), in contrast to the digital-media campus or Silicon clusters that focus on co-production, rather than self-promotion.

The branded media centre in city centre locations is therefore a familiar one in aspiring world and creative cities (Evans 2009c) with a common international property rationale driving them. For example, Berlin's Potsdammer Platz was always a symbolic centre of this city, co-opted during WWII for the Nazi People's Court, then a bomb site and no man's land of the Berlin Wall which further destroyed

Figure 6.2 Outernet, Tottenham Court Road, London
Source: Photograph by the author.

remaining buildings. Following reunification, Berlin was literally a hot property for development and for artists. In the early 1990s, the site was acquired by the Sony Corporation who financed this eight building complex including a 4,000 m² vaulted roof covering these buildings and housing Cinestar cinema, Sony Center and HQ and an IMAX theatre, which were used for Berlin's film festivals until these cinemas were closed by new owners in 2019. The centre has changed hands several times and is essentially a property investment vehicle. The latest owners plan a major refurbishment of the Sony Center Campus with an emphasis on health and well-being, and in their property-speak, *world class retail and future-focused amenity*. To what extent this future will embrace digital production and experience is still an open question, despite the digital determinism at large.

So that whilst the late-1980s/1990s were associated with the notion of the *Creative City* (Landry and Bianchini 1995; Evans 2017) to be populated in the 2000s with the mobile *Creative Class* (Florida 2005; Evans 2011), the *Digital Age* has coincided with – rather than driven – the so-called *Experience Age*, or what economic geographer Alan Scott has termed 'cognitive-cultural capitalism' (2014). In its atomised form, essentially visual (Jay 2002), and enhanced through VR and other immersive capabilities within various devices and apps, but in its enduring physical form, still retaining the appeal – to both artists/producers and audiences – of place-based collective engagement and 'spatial embeddedness'. This contradictory take on the experience economy (Pine and Gilmore 1998) also means, on the one hand, that digital cultural producers large and small retain many of the work styles and location preferences of previous cultural and other forms of industrial production (with the exception perhaps of amateur social media producers) and maintain some physical and social connection to their host communities, rather than a disconnection that Castells foresaw through the forces of global information technology (1996).

On the other hand, consumers have a confusing and uneven array of alternatives in access to cultural activities, via streaming and apps, but live 'authentic' experience and amenity are still alive and in demand (Nichols Clark 2011), whether in a theatre, festival, club or gallery, and in its global dimension, through the visitor economy – home and away (MacCannell 1996; Richards 1996). Designing-in the experience in some causal manner through the digital simulacrum also risks 'denying users routine creativity and precludes interpretations of "use" as a dynamic, ongoing achievement' (Redström 2006; Shove et al. 2007, 131), as if experience is a new concept and emotional engagement with art is limited and in need of (artificial) animation. As Chapman observes: 'it would be ridiculous to assume that material culture was somehow devoid of qualitative user experiences prior to the advent of experience design . . . indeed the design of experiences isn't any newer than the recognition of experiences' (2011, 92; see also Shedroff 2001). Another flipside to the Artificial Intelligence (AI) phenomenon has been illustration and image generation, drawing on databanks of images (not just those in the public domain) – in less than a minute after inputting some text prompts, an AI-image is returned to sender. No copyright or acknowledgement to the originators whose work has been 'scraped':

> these programs rely entirely on the pirated intellectual property of countless working artists, photographers, illustrators and other rights holders. . . . AI doesn't look at art and create its own. It samples everyone's then mashes it into something else.
>
> (Shaffi 2023, 8)

Digital experience, as distinct from cultural space, seems to presume that exaggeration and extra stimulation are required in order to capture an equivalent authentic experience, as if our oral and visual culture and tradition are also in need of digital enhancement and validation. In its current state (of development and governance), the digital is primarily a tool of the content industry's incessant thirst for filling the bottomless void of cyberspace – in a multichannel, multi-platform, multi-device world, as well as enabling access whether streaming 'live' theatre or music performances to local cinemas, or supplementing library provision through e-book lending services. With the advent of the next generation AI however – on the cusp of becoming out of control (with warnings emanating from the creators/ owners of this technology themselves, *vis* Musk, Google, Microsoft) – artificial and derivative content increasingly threatens to corrupt established forms of conceived *and* perceived space. Cultural spaces can therefore serve as both an antithesis to these anonymous and place-less digital spaces, but can also mediate between the real and the cyber space, whether through the agency of users, consumers or cultural producers. In this respect, the digital must remain subsidiary to cultural space, and not vice versa.

*

Creative digital is very much an evolution and combination of established arts and cultural production and manipulation (e.g. VR/AI data mining) and new technology delivery and dissemination platforms – currently heavily dependent on cultural content. As well as the more ubiquitous communication revolution, digital technology has also transformed much creative production and making, not just in architecture and graphic art practice but in the design and manufacturing process of products, including clothes. Fashion production and consumption spaces have not surprisingly been influenced by digital technology, and this is also manifested spatially through digital cultural spaces. Fashion production space is therefore considered in the next chapter.

Notes

1 www.punchdrunk.com/about-us/
2 For example, see https://holition.com/work/dunhill-holographic-fashion-show
3 The Lightroom occupies lower floors on the site of Meta's (Facebook, etc.) new London HQ.
4 www.storyfutures.com/https://xrstories.co.uk/about/
5 The Chinese-owned (ByteDance) TikTok, and its powerful, responsive algorithm, overtook YouTube as the most popular (by users = over 1 billion) social media platform in 2021.
6 The 'Creative Digital' firm is a blend, fusing technology and design in the context of Advertising, Communications and Marketing services alongside applications (mobiles) and (computer) games. Typically, it offers a number of products and services simultaneously to clients – a hybrid of creative strategy development, campaign implementation and technological development (Cities Institute 2010).
7 In the summer of 2007, a year before the collapse of investment bank Lehman Brothers, the idea of an emergent digital cluster around a nondescript roundabout on the City Fringe's eastern border developed. Mark Biddulph, the technical director of a social network travel start-up (Dopplr), broadcasts his observation over the net that there was an 'ever-growing community of fun startups in London's Old Street area' – a mix of innovative dot.coms, business software companies, established web-designers and a design-led digital printer. With tongue-in-cheek homage to Silicon Valley, he renamed the area Silicon Roundabout (Foord 2013, 54).
8 Richard Kastelein, 'Google launches facility in East London's Tech City' April 4, 2012 (https://connect.innovateuk.org/web/convergence/articles/-/blogs/googlelaunches-facility-in-east-london-s-tech-city;jsessionid=A7EE34E5CA3A199990FB8D46B5 38B113.MekushUdbew4

7 Fashion Spaces

Fashion, from *street* to *catwalk*, has particular place associations not least with the cities of fashion that represent a dominant hierarchy of design/designer-labels and luxury brands, which are of course also highly personalised and hard branded with their founders (Chanel, Gucci, McQueen, Prada, Dior et al.). Breward and Gilbert (2006) assert that the complex relationship between the fashion industry and urban growth underpinning contemporary understandings of global fashion is orchestrated around a fluid network of world fashion capitals, particularly Paris, New York, London, Milan and Tokyo, with the largest and most influential fashion companies considering these cities as 'vital, aesthetically challenging, entrepreneurial and central to business' (Chiles and Russo 2008, 7). 'Second-tier' and aspiring fashion cities also supplement this hierarchy such as Moscow, Vienna, Berlin, São Paulo, Kuwait City, Cape Town, Barcelona, Antwerp, Delhi, Melbourne, Sydney, Shanghai, Hong Kong, Mumbai and others aspiring to secure fashion city status, at least in their own region (Eicher 2010). As an 'old' cultural form, fashion retains high valorisation, even against the new tech sectors that supposedly drive economic growth and innovation. An indication of this is seen in the ranking of the richest person in the world which at January 2023 was Frenchman Bernard Arnault, the chief executive of French luxury goods conglomerate LVMH, which owns fashion labels Louis Vuitton and Dior among others, overtaking the Tesla/Twitter majority-owner, Elon Musk of the USA. The evolution of ownership of the great fashion houses in the twentieth century is, however, characterised by decreasing control by the founder/family: 'there is no longer a financial identification between couturier-founder and ownership' (Santagata 2004, 88), and as founders pass away (e.g. Chanel, McQueen), the brand and stock market value become synonymous.

Fashion brands and place brands find mutual advantages in both everyday and spectacular spaces. The live experience, as discussed in the previous chapter, still maintains its allure to trade, consumers and media alike and fuels fashion content and celebrity in all its guises. With the big four fashion weeks of Milan, London, New York and Paris now spread across the year (London hosts four fashion weeks, New York six), over 100 fashion week events are held worldwide, from Fiji to Panama, and from plus size and ecologically focused themes, to beach-wear in Goa, India,[1] a case of *mondialisation* also seen in the spread of Art biennales (Evans 2011). I use this term as the stage *before* the cliche *globalisation*, since

DOI: 10.4324/9781003216537-7

local and regional identity and production are still retained in many of these examples, even if the Fashion Week format is adopted in order to stake a claim within this international market and mileu (e.g. *Africa Fashion* exhibition at the V&A in 2023 and sessions on Africa's 'Premier Fashion Week' in Lagos). The return to catwalk shows, after the enforced closure as a result of the Covid-19 pandemic and travel restrictions, saw the resurgence of extravagant displays from fashion houses which combine theatre, film and online hybrid versions for those not on invite-only lists. For example, Dior's 2021 catwalk show was held at the 70,000 seat Panathenaic stadium in Athens for a guest list of only 400, backlit with fireworks and a full orchestra, as fashion journalist Carter-Morley observes: 'catwalk shows are symbolic of fashion's identity as a creative art as well as a business' (2021, 21). Although an international road show, in this case local Greek makers were also commissioned to provide woven pieces using traditional Greek jacquard techniques. Gucci's 2023 show also sought to seal Seoul's status as a fashion city powerhouse celebrating its K-pop and street culture, and a burgeoning luxury goods market (South Koreans spend an average £260 on luxury goods each year, the highest per capita in the world) with a catwalk extravaganza held at the Gyeongbokgung Palace, the first fashion show to be held at its ceremonial courtyard. Guest of honour at the event was brand ambassador Hanni of the appropriately named girl band NewJeans. Another LVMH luxury brand Louis Vuitton's catwalks have been staged at the Miho Museum, Japan, and the Niteroi Gallery, Brazil, and again in Seoul, along the Jamsugyo Bridge, whilst in London, new designers took over a cavernous former Selfridges car park and the Olympic Aquatics Centre, so that despite the opportunities from Virtual Reality and CGI such as avatars or 'digital mannequins', high-end fashion looks to authentic cultural and everyday spaces to perennially frame itself.

More recently however, Breward and Gilbert (2010) observe that this geographical fashion city hierarchy is changing, as the system of fast fashion with its reliance on overseas producers disrupts the traditional relationship between time, place and fashion creativity (Ottati 2014), a shift accelerated by online fashion retailing and social media (*viz.,* Instagram and TikTok 'influencers'). The production of garments has, of course, been fragmented by some of the early examples of borderless globalisation through outsourcing and offshore production that has broken more vernacular regional production and consumption chains, such that few fashion brands manufacture in the regions in which their design houses and key consumer markets reside (although in the latter case this is changing as China and other Asian consumers discover western brands, both luxury and fake as well as high street via designer outlets). This has generated complex production and supply chains largely hidden from the consumer's view – from design, raw materials (e.g. cotton, wool), processing and finishing, to retail and showcasing – further complicated by online consumption through e-commerce and physical-virtual hybrid retail (Alexander and Alvarado 2017). The pursuit of ever-lower cost production has seen countries gain then lose their place in this chain, for example, Portugal and Turkey, to countries such as Vietnam and China. The advent of sustainable development and fair trade has however begun to lift the lid on the social and environmental impacts of these practices with responses in the form of sustainable design and production, re-onshoring of textile manufacturing, farming (e.g. Merino sheep), if not growing (e.g. cotton in Manchester, once known as *Cottonopolis*) and waste recycling, as well as make-do-and-mend/repair spaces (Evans 2018c). New advances in block chain technology have also made it easier and more transparent to source and validate supply chains and identify originators (reducing the ability to copy designs), particularly useful where clothing labels do not reflect or credit their true origins (Braddock, Clarke and Harris 2012) – for example luxury brands manufactured in Shanghai but supposedly 'made' in Italy, or made by Chinese, but *in* Italy.

Clothing is perhaps the most personal cultural product – a *second skin* – and whilst some culture can be experienced in private and tastes not necessarily shared (i.e. listening to music), fashion and its accessories (e.g. jewellery) are also a form of display and identity, even when primarily utilitarian:

> Clothing is a status-conferring good. Most forms of apparel above the commodity category, and even some apparel within that lowest-level category, function as what economists call 'positional goods' whose value is closely tied to the perception that they are valued by others.
>
> (Raustiala and Sprigman 2006, 1718)

As a material-based cultural practice, clothing is one of the most traditional and pervasive cultural firms – and as Sennett observed (1997), sight and appearance triumph in the modern world, which Gonzales attributes to a combination of structurally modern space and historic cultural ideals (2010). Fashion also reflects the socially constructed high-art/popular-culture dialectic present in other art/creative forms, including the extremes of craft/hand-made and mass production/fast fashion, and other key distinctions. For example, in the classification of creative industries, 'designer fashion' is included in government definitions and measurement of the creative economy and employment, whilst other fashion production and activity are not (DCMS 2001; Bakhshi, Frey and Osborne 2015). Attributing 'creative or non-creative' to economic activity can be a reductive venture: 'rather like unpacking a Russian doll; once the layers are discarded ay heart it appears an amorphous entity' (Galloway and Dunlop 2007, 29). Spatially and culturally, the relationship between eponymous sweat shops and between street/pop culture and *haute couture* represented in design houses, upmarket fashion retail and museums, is, however, closer than is often realised.

From Street to Catwalk

Cultural industries, such as film, music and publishing, are increasingly concentrated, characterised by a small number of firms that produce a large share of total industry output, although this oligopoly is diluting as the digital economy spreads – but not of course in the tech sector itself. On the other hand, the degree of concentration in the fashion industry as a whole is relatively low, with a large number of firms of varying size and producing and marketing original designs, although concentration through acquisitions is evident in luxury-brand holdings (e.g. the French LVMH and Kering). No single firm, or small group of firms, represents a significant share of total industry output. Since the late 1980s, however, as luxury evolved from privately owned family-run houses to publicly traded global conglomerates, couture has also become an exercise in branding to sell logo-embossed lipsticks, perfumes and handbags to the masses (Thomas 2004): 'everything from baby clothes to disposable cameras slapped with [Tommy Hilfiger's] familiar emblem' (Wolfson 2023).

Fashion, like other highly commodified cultural forms, has created and sought to maintain market distinctions (Bourdieu 1984), characterised by mediated categorisations, from *Haute couture*, *Prêt-a-Porter*/Ready-to-Wear, to High Street/ Everyday, *Street*-wear, Fast Fashion and Second-hand/Vintage, but with opaque overlap between these otherwise brand/price-distinguished products. This hierarchy (modelled on Maslow's Hierarchy of Need – and see Doeringer and Crean 2006; Raustiala and Sprigman 2006) has held good in the modern era (Figure 7.1), but so-called luxury brands increasingly bridge both middle and fast fashion sectors, as well as appropriate street and popular culture in order to widen their market reach and turnover and to extend brand identity. Fashion spaces occupied and manufactured in this process both reflect and transcend these distinctions.

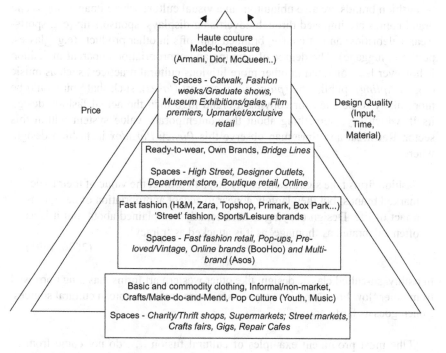

Figure 7.1 Hierarchy of Fashion and Fashion Space

Haute couture has a particular symbiotic/sycophantic relationship with museums and with media such as film (premieres, *red carpets*), since outside of perennial catwalk shows and associated events and their press coverage, the public has little contact or experience of high fashion beyond a few upmarket shops in major cities. Consumers and fashion voyeurs – practicing what in previous times would be called 'window shopping' – experience firsthand via high street and ready-to-wear stages of the production chain, where near-copies or sub-branded versions of haute couture are made available to a mass market, a process further replicated through fast fashion and designer-replicas both legal and counterfeit. Many fashion design firms operate at multiple levels of the pyramid. For example, Giorgio Armani produces couture apparel, a premium ready-to-wear collection marketed via its Giorgio Armani label, differentiated bridge lines marketed via its Armani Collezioni and Emporio Armani brands, and a 'better clothing' line distributed in shopping malls via its Armani Exchange brand. The borders between product categories are increasingly indistinct as so-called haute couture meets fast fashion retail:

> On a frisky Friday night in September, Paris's chic and hip flocked to the rooftop restaurant of the Centre Pompidou to celebrate fashion's coolest new collaboration: Karl Lagerfeld [Chanel and own-brand designer] and H&M where Lagerfield unveiled his new clothes for the Swedish retail chain.
>
> (Thomas 2004)

Fashion brands are also ubiquitous in a visual culture where images, logos and brand names are imposed through advertising displays, sponsorship (e.g. sportswear, celebrities) and of course, branded spin-offs in other products (e.g. glasses, perfume, luggage). The degree of copying and appropriation apparent in fashion is however less tolerated in other more litigious cultural practices, such as music (e.g. *sampling*), publishing (*plagiarism*) and art (*fakes*), such that you would be more likely to be sued for copying a brand logo than the actual fashion design itself, which says something about the contemporary value system within this sector. Raustiala and Sprigman observe this *Piracy Paradox* in fashion design, where:

> Fashion firms take significant, costly steps to protect the value of their trademarked brands, but they largely appear to accept appropriation of designs as a fact of life. Design copying is occasionally complained about, but it is as often celebrated as 'homage' as it is attacked as 'piracy'.
>
> (2006, 1691)

In reality, so-called classic design, like other western art forms, has long borrowed from other 'low brow' and in many senses, more creative protean cultural sources. Cohen goes as far to argue:

> [T]he most prominent examples of cultural fusion . . . do not come from global centres, but rather from the world's periphery; they represent primarily

an attempt at localisation of global stylistic trends – the fusion of Western artistic styles or forms with local third or fourth world cultural elements.

(1999, 45)

However, it is the value-added of the brand-logo attached to everyday fashion items that secures its status, as founder of Bathing Ape 'streetwear' brand (not to be confused with the actual 'street') comments: 'when I first started out it was almost shocking that the fashion world would pay attention to street clothing. Now that's normal' (Benigson 2014).[2] So in Chanel's Spring Paris catwalk show that year, each model was wore trainers, of course branded, and customised and decorated with jewels and embroidery. In the USA, the eponymous clothes brand, Tommy Hillfinger, epitomises the primacy of brand-led promotion and identity (Hilfiger actually admits that he's more excited by his brand as a marketing and communication exercise than he is about fashion designing). Inspired by Andy Warhol, who he met in the early-1980s at his Factory, and the power of symbolism, Hillfinger first studied symbols such as the Rolling Stones tongue, the Nike swoosh and Mercedes star, as well as Chanel 'Cs' and Gucci 'Gs'. Appropriating pre-existing sportswear – rugby shirts, chinos, sailing jackets and a preppy private school style, he logoed these symbolic if originally utilitarian items, but it was the take-up of these by young black Americans, including rappers in the early-1990s that secured the brand's dual aspirational status, worn both on the street and by rich white teenagers – 'young street kids and rappers wanted to wear the clothes because they wanted to look rich' (Wolfson 2023) – and rich kids to look cool. Whilst numerous wealthy black pop music stars have branded their own fashion labels and in some cases joined fashion house design teams (e.g. Pharel Williams to Louis Vuitton), street design itself – akin to street art (Chapter 8) – is largely appropriated by established fashion industry and brands without licence.

When fashion was more a question of class, the bourgeoisie imitated the aristocrats. In the twentieth century, however, the models no longer came from the aristocracy. In 1985, the top 21 houses classified in France as *haute couture* employed only 2,000 workers in their ateliers and had no more than 3,000 female customers worldwide. However, these designer labels no longer made a profit but did so from selling more standardised *pret et porter* clothing, accessories, perfumes, leather goods, tableware, pens and so on to a wide market. So that, by the mid-1980s, Yves Saint-Laurent derived two-thirds of its profits from royalties and Pierre Cardin relied on over 600 foreign licences (Sassoon 2006, 1367): 'Haute Couture had become a marketing tool to sell to a mass market' (Lipovetsky 2002, 89). From the 1950s onwards, fashion has clearly moved from being imposed from above to spreading from below (Crane 2000, 14), and although there is the need to use celebrities as transmitters of fashion trends, the mimicry remains (Gonzales 2010). However, as Simmel himself pointed out (1971), mimicry cannot be complete because both society and fashion demand differentiation. So, for example, although the bourgeoisie did imitate the aristocrats, they always introduced a new fashion when the earlier one had filtered down to the lower classes, preserving social difference. So that when a particular fashion style has been more or less accepted by the mainstream, those consumers with agency and a stake in the process differentiate themselves again by adopting new styles and innovations (Gonzales 2010). This is also a key appeal of fast fashion to *haute couture*, epitomised by Spanish fashion brand Zara – the fast turnaround enabled by digital design and manufacture (and online retail/social media), with the shelf life of fashion items literally reduced to between four and six weeks.

Fashioning Museums

Fashion as material culture has also gained growing attention in museums, as well as through placemaking around fashion districts and events. Whilst costumes and historic artefacts have long featured in museum collections – from folk art and national dress to imperial courts – dedicated fashion museums have been opened to capture popular interest and scholarly work in contemporary fashion and styles and fashion icons, as well as fashion eras and places (e.g. Carnaby Street, *Swinging London* – Breward 2004). Fashion's relationship with other popular culture, notably music (e.g. David Bowie[3]), theatre/dance (e.g. Diaghilev) and art (e.g. Freda Khalo), has provided a rich seam for curators and museum programmers. Living fashion designers are frequently celebrated in major exhibitions at the V&A in London and Metropolitan Museum of Art (Met), New York, with retrospectives producing some of the most well-attended exhibitions for these venues, for example, *Dior* (600,000 visitors), *McQueen* (500,000) at the V&A, and *Heavenly Bodies: Fashion and the Catholic Imagination* (1.6 m visitors) featuring work by designers Chanel, Lagerfeld and Dolce and Gabana, and outstripping the Met's 'blockbuster' Picasso, Mona Lisa, Impressionists and Tutankhamun exhibitions. The Met of course hosts the annual Gala, billed as 'fashion's biggest night out', an ostensible fundraising event for the museum itself (sponsored by fashion brands), and attended by a mix of film, pop and sports stars alongside fashion industry glitterati or *fashionistas*. The reality, however, is that much fashion production remains hidden, not just the uncomfortable truth of sweatshop manufacturing, both offshore and in major fashion cities, but also where the contribution of migrant communities, from Hugenots (e.g. silk-weaving) to Jewish tailors and garment firms, is understated, as Whitmore remarks: 'few recognise the influence Jewish designers and makers had at all levels of the fashion trade, from establishing the ready-to-wear industry to dominating fashion meccas like Carnaby Street in the 1960s' (Addley 2023, 13). An estimated 60%–70% of Jewish immigrants to London in the early 1900s worked in the 'rag trade' (a similar proportion in New York also worked in the garment trade), to be followed by Bangladeshi and Pakistani migrants in the 1950s, as a successful Brick Lane restaurateur, who started his working life in a Brick Lane clothing factory recalls: 'we used to do a lot for Debenhams and Selfridges clothing. Well, that's the tradition: East End production, West End retail'.[4] An exhibition at the Museum of London Docklands celebrates this cultural contribution to the Fashion City: *How Jewish Londoners Shaped Global Style*, although iconic fashion items made by Jewish firms and designers for film, music and sports stars and royalty, are yet to be located, with few held or attributed in existing museum collections. In the USA of course, several fashion design icons were Jewish – from Levi Strauss and Ralph Lauren to Calvin Klein.

The museumification of modern fashion can also be viewed as the aspiration and desire to be taken seriously as an art form, at a time when cultural branding has been adopted by museums and fashion labels alike. Major exhibitions at the V&A subsequently tour around the world furthering the cultural diplomacy mission of its foundation following the Great Exhibition of 1851 (Evans 2020c and Chapter 3). Today, curators serve in both university art schools and museums, such as the V&A and Tate Gallery, and benefit from fashion company sponsorship of exhibitions, research and archives. This is not surprising given the success of fashion-brand museum exhibitions and the source of future museum collections. An early indication of the experiential revolution (Pine and Gilmore 1998), where 'retailers must now also deliver in-store experience to their target audience' (Servais, Quartier and Vanrie 2022) is seen in major fashion store make-overs. The synthesis of culture and consumption reached new heights on the site of the former Guggenheim museum in New York's SoHo district where a new Prada store designed by Dutch architect Rem Koolhaas presaged new stores for Tokyo and San Francisco – all part of the 'Prada Universe'. According to Koolhaas, this will lead to the reshaping of the concept and function of shopping pleasure and communications outlets so as to fuse consumption and culture. Stores will play a dual role acting as entertainment and cultural venues, transforming into an evening theatre with drop-down ceiling to floor staging for performances (Evans 2003a, 434). The stores were conceived to be social laboratories that 'encourage interaction and exploration rather than mere consumption. Shoppers will become "researchers, students, patients, museum goers" in an environment that borrows elements of the theatre, trading floor, museum and the street' (Emerling 2001). This has not been realised however, with Prada and Koolhaas's hype and hubris undermined by the other revolution – online shopping – as well as competition from actual museums nearby (e.g. MoMA and museum retail outlets), and tragic events, notably the Twin Towers attack in September the same year that the Prada store opened.

In the twenty-first century, dedicated fashion museums have been founded in cities such as Antwerp (Pandolfi 2015) and Hasselt, both in Flemish Belgium (with the Flanders region a long-established clothmaking tradition) and also in London, where designer Zandra Rhodes opened the Fashion & Textiles Museum in 2003, in a colourful building designed by Mexican architect Ricardo Legorreta, which stands out along the drab, narrow streets of Bermondsey, South London (Figure 7.2).

This museum is now operated by a further education College in Newham, east London (in the absence of either a luxury brand or state sponsor). Newham is an area of east London that has undergone major regeneration, primarily as a result of the London 2012 Olympics and major retail-led development (e.g. Westfield Shopping Centre) and which boasts a growing Fashion District, to include the new London College of Fashion relocated in 2023 from its Oxford Street heartland to Stratford's East Bank as part of the *Olympicopolis* project (below).

Figure 7.2 Fashion and Textile Museum, London
Source: Photograph by the author.

Modern fashion museums based on former private collections have also opened since the 1960s, for example Fashion Museum Bath in west England, originally based on the collection of designer and collector Doris Langley Moore. The museum is currently closed for its planned move from the Assembly Rooms to the Old Post Office city centre site, with its archive relocating to Bath Spa University at Locksbrook. Most fashion collections outside of major museums and archives, including university and trade guild archives (e.g. handbags, milliners), reside in fashion house collections, from 'heritage' shoe brands Bata (Toronto, Canada) and Clarks (Somerset, UK), to luxury brand displays in villas and stylish mansions – Gucci, Ferragamo (Florence); Dior, Yves-Saint Laurent (Paris) and Balenciaga (Basque region, Spain), reinforcing their fashion city-region status (Jansson and Power 2010).

Fashion brand museums, essentially vanity projects of their founders, also represent the celebration of art and commerce, philanthropy and cultural validation, from LVMH's new museum in Paris (a former bowling alley), the Cartier Foundation outside Paris, to the Prada Foundation, Milan (built on the site of a distillery) (Figure 7.3) and which sees fit to ask: 'What is a cultural institution for? This is the central question of today. We embrace the idea that culture is deeply useful and necessary as well as attractive and engaging. Culture should help us with our everyday lives, and understand how we, and the world, are changing'. The world awaits the Tesla or Musk Museum and the founder's even more galactic vision. It is

Figure 7.3 LVMH, Paris and Prada, Milan Museums

Source: Photographs by the author.

interesting to note however that these fashion company foundations display *art*, not clothing. The LVMH museum was designed by Frank Gehry to resemble a tilting ship with 12 sail façades, but it is located in a former forest the *Bois de Boulogne*, adjoining a children's amusement park, whereas Gehry maintained, with no sense of irony, 'nobody wanted an intrusion in the park' (2014, 257).

Fashion Districts

Whilst fashion cities are identified with their international status combining both endogenous and imported/exported talent, brands and markets, fashion production is more localised, as with many other cultural industries. With a focus on the workshop-atelier, this is the place where designers 'decipher their working environment through their fantasy and imagination . . . and their view of society and history, realising a unique link' (Chiles and Russo 2008, 5; Maramotti 2000). Fashion products can thus be described as having idiosyncratic character and affecting economic behaviour in two ways: 'by means of involvement in a community or social group and by immersion in the productive atmosphere of the cultural industrial district' (Santagata 2004, 79). Fashion districts represent particular clusters of design and manufacturing activity and associated retail and showcasing. This is seldom monocentric with the various stages in material processing, manufacturing, finishing, intermediary wholesaling and retailing, as well as design and innovation, separated by scale and distinctions (boutique, designer, upmarket, mass produced/

retail, etc.) and spread across polycentric clusters, with either loose and minimal ties or more often forming strong production chain links. Add in the influence of art and design institutions, museums, street markets and tourism, and fashion districts represent only one element in the fashion creative production chain, in practice focused on small enterprises, designers and more craft-oriented production, whether rooted in traditional makers as in Emilia-Romagna, northern Italy, or as part of new economic development built around the regeneration of post-industrial areas with fashion and textiles heritage. Case studies and literature on fashion districts thus feature in work on industrial, cultural and creative clusters (Gong and Hassink 2017); on regeneration and place/city branding (Kavaratzis, Warnaby and Ashworth 2014; and on heritage quarters (Shorthouse 2004), illustrating the importance of this activity to both symbolic and political economies (Zukin 1995).

New York's Garment District, as analysed by Norma Rantisi (2002), for example, highlights the polycentric nature of production and innovation (Figure 7.4) with low cost/sweatshop fast fashion symbiotically co-existing with *haute couture*, up-town retail and brands, alongside art and design institutions and showcasing, as well as co-location with a range of cultural and commercial enterprises all in close proximity, which complement and complete a particular opaque production chain.

Los Angeles, second only to New York in fashion industry concentration, also benefits from its association and proximity to Hollywood's film and TV sector. Together these two cities dominate the national fashion industry with 46% of wholesale, 50% of manufacturing and 43% of design, with the nearest rivals Miami and Chicago providing just 5% and 4% of the sector. Whilst the fashion system here, as in other post-industrial cities ('post' in this respect is a misnomer since industrial production is still apparent), has moved to a more design-oriented 'cognitive-cultural economy' (Scott 2014), 'production is also important providing the materials and supplies necessary for the innovation and preliminary production processes . . . linked to the need for fashion to have just-in-time materials and production for samples and high-end design collections' (Williams and Currid-Halkett 2011, 3043; Rantisi 2002)

This dual system of fashion production – and consumption – is also evident in London's fashion quarters. Far enough apart to avoid planning integration, consumer collaboration or visual coherence, the so-called sweat shops of inner and outer east London have no ostensible links to the design houses, upmarket shops and fashion institutions of the West End and central London. However, as independent designers and production have moved east, following an accumulation of area-based regeneration, creative clustering and a growing disparity in the cost of both housing and workspace (Chapter 5), the fashion design process and manufacturing supply chain has become more spatially and culturally explicit. Whilst these sweat shops had always supplied the West End and just-in-time catwalks, they have now been joined by independent designers, student start-ups and new graduates, as well as boutiques, event spaces/galleries and new workspaces and facilities such as digital design and manufacturing. The false distinction between *designer* and *ordinary* fashion is illustrated by the maps of these classified firms located in the City Fringe where co-location is evident (Figure 7.5), with this established cluster both connecting and elevating a nascent fashion district further east.

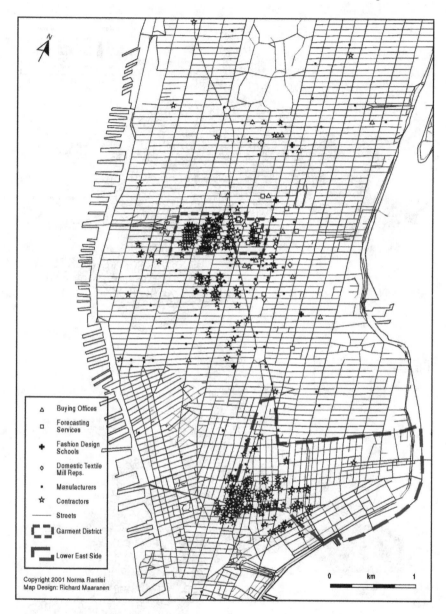

Figure 7.4 New York City Fashion Design Production System

Source: Rantisi (2002, 595).

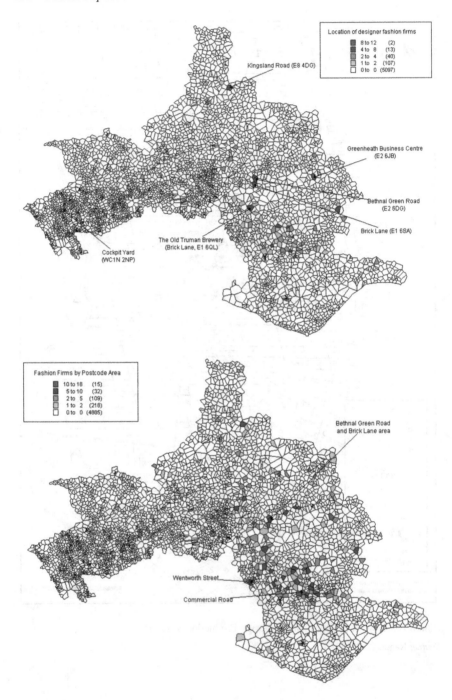

Figure 7.5 Co-Location of Designer Fashion and Fashion Firms in the City Fringe

At the start of 2016, 36,200 people were directly employed in fashion retail, design, manufacture, distribution and advertising in this area – an increase of over 10,900 jobs since 2010. East London is also home to over a quarter of the capital's fashion enterprises and employment and is therefore driving the growth of London's fashion design, retail and manufacturing sectors. The significance of the fashion cluster had been identified ten years ago with the designation of the City Fringe area bridging the edge of the City of London and East London as one of the city's Creative Hubs – a spatial strategy promoted by the then Mayor's London Development Agency *Creative London* (Foord 2013; LDA 2003). *Designer Fashion* and (ordinary) *Fashion* therefore co-locate/co-exist with a range of other cultural and creative industries, clustered around workspaces, exhibition/showcase facilities, such as the Truman Brewery, and specialty retail streets, such Brick Lane, Bethnal Green as well as ordinary neighbourhood retail and street markets. Meanwhile, the West End, centred on the retail and touristic promenade of Oxford Street continues a downward trend, accelerated by the general decline in high streets from online and out-of-town shopping, the Covid-19 pandemic, and critically, the impending loss of its iconic fashion institution, the London College of Fashion (University of the Arts London). This college moved to its brand-new building in 2023 in Stratford, East London (Figure 7.6), as part of the East Bank development (formerly known as *Olympicopolis*) on the parkland site of the London 2012 Olympics and as part of a self-styled fashion district nearby (see Chapter 3).

The East London Fashion District (essentially a rebranding of a residual manufacturing and designer-making community) can be seen as a combination of area-based regeneration, regional levelling up[5] and a drift eastwards to an area that was a vibrant nineteenth-century industrial zone and source of innovation, for example in materials (e.g. Parksine, an early plastic) and armoury/ordnance (e.g. Enfield rifle). Today this wider area, known as the Lower Lea Valley, already contains more textile firms than other UK regions, where the urban imaginary is typically identified with large-scale manufacturing in the North of the country, notably in Yorkshire and Manchester, North West England (RSA/UCL 2018). Like other cities, this largely hidden manufacturing and processing cluster serves a wider and central city fashion ecosystem, effectively a swathe linking mainstream fashion

Figure 7.6 London College of Fashion Oxford Street and East Bank Stratford
Source: Photographs by the author.

brands, stores, small workshops and designers and larger-scale manufacturing. To this extent, designating a small area as the Fashion District tends to understate this relationship, but conforms to a political desire to focus attention and investment in one, more visible and visual part of a creative production chain (BOP 2016; Vecchi and Evans 2018), where it plays a part in a wider scalar narrative of regeneration (Gonzales 2006).

The strategic relocation of major cultural institutions and facilities driven by state regeneration, regionalism and associated incentives also raises important questions for how cultural places are considered and created. Examples such as the move of the BBC to Salford in the North West, from White City in West London; Danish broadcaster DR and IT University relocated from central Copenhagen to the new satellite city-extension of Ørestad (City of Copenhagen 2003); and the con-solidation of art and design colleges in moves to new regeneration sites such Pobra Fabra University to Poblenou in Barcelona and Helsinki Art & Design University to *Arabianranta* industrial districts (see below), all provide cases of what, and how long, it takes for these new sites of cultural production to become recognised and connected to local cultural ecosystems, and the extent to which they achieve the same level of symbolic and cultural value of their original locations. Little research or consideration is however placed on these former sites, the impact on their local cultural economies and everyday life and the loss and damage incurred as a result of enforced moves (e.g. staff not able/willing to move location). In the case of White City in London, originally arable farmland, this had been the site of the 1908 Olympics and successive Great Exhibitions (forerunners of International EXPOs – Chapter 3) between 1908 and 1914, only to be interrupted by World War One. The BBC Centre was built here in 1990 on the site of the White City Stadium, a legacy from 1908 Games, and dismantled in 1984/5, housing the BBC Television Centre and associated facilities, including BBC radio studios. The move of the Centre and most other studios to *MediaCityUK*, Salford, in 2011 saw the loss not just of direct employment, but also a wide range of secondary employment and suppliers, including fashion/accessories, costume and set designers and makers – to dog han-dlers. Much of this activity moved (some, but not all BBC staff) or was supplied in the new location, but the net gain in employment was minimal. What little evalua-tion of such cultural relocation that has been carried out, has been limited to media/ broadcasting and IT activities, ignoring wider production chains from fashion and other crafts employment (Scott 1997; Rantisi 2002)[6] and the multi-layered history embedded in this area. New occupiers of the White City sites include Soho House, a private members club for 'creatives' (ironic, given that most of the 'creatives' have moved out), and a new faculty of Imperial College, not one engaged in cul-tural or creative fields, but chemistry. Another example is the Orestad extension to Copenhagen, with a linear array of modern architectural offices and retail blocks, new schools and housing, and the relocated Technology University and Danish broadcasting facility. Here, the retro-fitting of cultural animation and activity more associated with run-down buildings and areas is sought by the developers, but who admit that: 'smooth planning of infrastructure does not do the trick of creating real life in new built dwellings'. Not surprisingly artists and designers were reluctant to locate in this new territory (Evans 2009b) and Danish fashion houses, design centre and colleges maintain their presence in downtown and historic showcase sites (e.g. City Hall).

Former (or redundant) fashion and textile districts that have retained much of their architectural heritage have also been prime areas for both gentrification and renewal, building on their historic associations and environment. Whilst many of these areas have been re-occupied by higher-value creative industries and advanced producer services (e.g. media, architecture and design studios), as well as housing and retail activity, such as the Lace Market, Nottingham (Shorthouse 2004), others have emerged from structural change in larger-scale, regional manufacturing (including textiles), for example, Barcelona's Poblenou district to reinvention in the new fashion economy as the 'catwalk of the Mediterranean' (Chiles and Russo 2008) and Helsinki's *Arabianranta* university-creative industries district developed on the former Arabia ceramics factory complex. (Evans 2009b). Elsewhere, smaller districts have been developed from less substantial beginnings. By definition, these examples are more place-bound and host smaller-scale forms of fashion, linked to independent designer-makers who are often graduates from local arts academies (e.g. Antwerp, Maastricht). These enterprises and designers are dependent upon social networks and urban milieus in which they operate (Aage and Belussi 2008), and the more active involvement of the local public with fashion blogging, co-design or customisation of clothing, and practices of shared consumption. For instance, in the case of Arnhem, the Netherlands, essentially an urban regeneration project in the deprived working-class neighbourhood of Klarendal, the city had existing fashion roots via the Fashion Design department of the ArtEZ Institute of the Arts and successful fashion designer alumni, as well as regional HQs of fashion retail and brands in the city. More than 50 jobs were created in the creative sector since 2005, with a sample workshop that can organise small-scale production of up to 150 pieces; Arnhem Fashion Connection – aiming at a closer co-operation between ArtEZ's fashion department and the vocational training institute RijnIJssel; as well as Fashion shows/Fashion Night events, and even a fashion and design hotel, Modez, with 20 rooms all decorated by fashion and product designers educated in Arnhem (Jacobs 2014).

Fashion Heritage

In an unapologetic world fashion city, Milan, the Ticenese fashion quarter, a picturesque canal-side cluster, has to date survived in the face of gentrification and wider city development of its transport infrastructure and mega-projects, notably the Milan 2015 EXPO (see Chapter 3). In a city of fashion, and of design (furniture/interiors) – celebrated in the annual *Design Fiera* – attempts at establishing major *grand projets* such as the city-centre *City of Fashion* and *World Jewellery Centre* were resisted by fashion designers and others, who suspected the schemes were primarily property-based (Bolognesi 2005) and have repeatedly failed as result (Pandolfi 2015). (Gucci's HQ is located in a warehouse area in a former factory building, others are located outside of city e.g. Aspesi in Legnano and Gucci in Tuscany). However, a long-established familial and trust-based system in the Ticenese fashion district has managed to adapt to both market and political change:

> a dual local-global orientation [which] characterizes the virtuous circle of pro-
> duction-consumption . . . [where] a part of the goods and services produced in
> the quarter are locally consumed by those who live and work there; another
> part is locally consumed by the transient population; another part is exported.
> Cultural entrepreneurs work on the transformation of meanings to resell them,
> transporting them from one context to another, adapting them to a diverse public.
>
> (Bovone 2005, 377)

This urban workspace is clearly inhabited and lived every day and able to maintain a place authenticity that satisfies the perception of different users at different times – that is, 'differential space'.

The image of a material-based cultural product, but one distinguished by brand and media-driven fashion, would suggest that its fragmented production and fickle consumer habits precludes continuity and place, even from among the luxury brands that trade on authenticity, but exploit through low cost/poor conditions (human, environmental) in footloose manufacturing and retailing/distribution. However, where rooted in localities, fashion production can therefore endure not only in village-based making familiar in small island states and less urbanised developing countries (Evans 2000) but also in highly globalised and urbanised countries such as the UK. For example, the John Smedley enterprise was founded in 1784 at Lea Mills, Matlock, in Derbyshire by John Smedley and Peter Nightingale (Florence Nightingale's great-great Uncle). The pair were Inspired by Richard Arkwright who had pioneered the factory system 13 years earlier at the beginning of the industrial revolution a few miles away in Cromford. They set about building a spinning mill – Lea Mills was an ideal setting; the brook that runs through the village provided motive power and a constant source of clean running water. Today the Lea Mills factory is still John Smedley's home, making it the world's oldest manufacturing factory in continuous operation (Figure 7.7).

Figure 7.7 John Smedley Factory, Derbyshire

Source: Photographs by the author.

Initially, John Smedley produced cotton thread, not dissimilar from other mills in the Derwent Valley, before expanding production to underwear, vests, athletic trunks and long john's. During both World Wars, a major part of production was turned over to the war effort – mainly producing knitted underwear for the forces. In the late-1950s, the decision was made to begin manufacturing fine gauge outerwear for men and women. The iconic John Smedley Polo Shirt, crew, V- and Roll-necks for men (worn by Sean Connery as James Bond) and elegant twin sets and jumpers for women (*viz.,* Audrey Hepburn) were key pieces that projected John Smedley into the heart of British fashion – these garments still remain a very important and popular part of their collection. Most staff travel from within a ten-mile radius to work there, around 400, including a small design team, many related to one other and across generations. The firm is therefore embedded in the local community and environment on which it depends, with ownership still maintained across an extended family dating back to its founders. The factory, of course, operates a shop for visitors and online service but maintains stores across London and in Japan, its prime overseas export market, with digital channels opening up in China. Whilst design and production are rooted in place, its raw material is however imported – Sea Island Cotton from the USA (Long Island) and Merino wool from Australia, with intermediary finishing undertaken in Italy before final weaving and finishing in Derbyshire. It may take some time, however, for the effects of climate change to farm Merinos in Derbyshire with fine enough wool to meet Smedley's specifications.

Where material is sourced locally, production takes on a particular vernacular character. This is the case with the Burel factory in Manteiga, nestled in the valley of the *Serra da Estrela* of central Portugal. Founded in the nineteenth century, this factory was built on the wool used in shepherds' cloaks or coats made from *Burel*, a tightly woven wool fabric that kept them dry and warm. As with other European producers, Portugal's wool and textile industry was however impacted by lower cost producers and most factories in the area closed. It was an unlikely intervention from an eco-hotel developer that rescued a remaining factory and Victorian-era machinery in Manteiga (Figure 7.8), in time for the last generation of weavers and machinists to still be available to transfer their skills and knowledge to renovate and re-establish the factory for contemporary production. As well as machinery, cards and hand-written pattern books that produced the textile style were also recovered and were able to be used in both traditional and contemporary clothes and furnishings. Its limited-edition clothing range *Woolclopedia* claims to re-use and recycle all waste from its slow production process. The firm maintains dedicated shops in Lisbon and Porto, as well as decorating the associated local hotels who organise factory tours, and Burel is exported widely, including to major stores in London and New York. Its ongoing challenge is in attracting younger staff and transferring design and technical skills and knowledge from older generations who work 'post-retirement', a situation exacerbated by its rural location and according to factory staff, an over-generous state unemployment benefit system which does not incentivise work or on-the-job training.

Figure 7.8 Burel Textile Factory, Manteigas, Portugal

Source: Photograph by the author.

Fashion's Waste of Space

A downside of our obsession with *in-fashion* clothing novelty and image, which has been enabled by fast low-cost fashion supply and branding, is the sobering fact that the industry contributes 10% of global man-made carbon emissions, and clothing has been the fastest-growing waste stream over the past decade. Since the 1970s, textile waste has increased by 800%, and in the UK alone, 30 million items of clothing are sent to landfill every week. Due to its combination of different materials, including non-bio-degradable, synthetic and micro-fibres, waste authorities struggle with recycling and treatment, leaving households unable to dispose of clothing and textiles in their waste, or recycle them in refuse bins – only 1% of clothing is actually recycled as a result. Microfibres increasingly used in sports and leisure-wear, account for over a third of global microplastic pollution, whilst fashion produces 20% of all waste water – over 800 gallons alone are used to make one pair of jeans. The clothing waste cycle therefore presents a different, but still connected link in the fashion and material ecosystem. For example, the value of clothes in the average UK household is estimated at £4,000, 30% of which have not been worn for at least a year – as an indication of their surplus value, over 1,000 clothes banks (Figure 7.9) are stolen from car parks each year in the UK, to be 'rebranded' and sold to other unsuspecting charities (Evans 2018c).

A study commissioned by the supermarket Sainsbury's also found that three-quarters of consumers throw away rather than recycle or donate unwanted garments (Evans 2018c). Before this stage, nearly half of clothing bought online is returned to suppliers and retailers, costing the fashion industry £7 billion each year, and rising, plus unnecessary transport and environmental costs.[7] A significant amount of these returns are not resold, but are disposed of, with fast fashion's life cycle around six weeks before the product value becomes 'uneconomic' (15% of electrical goods returns suffer the same disposal fate).[8] The practice of burying waste in landfill sites had been a late-nineteenth-century phenomenon, up and until then human waste archaeology had remained largely unchanged – largely domestic and sacred items (e.g. coins, glass, pottery) with early landfill sites located on the coast in soft, sandy soil. Only industrial manufacturing which started to rely on non-degradable plastics and other oil-derived products and synthetic materials, for example, nylon and polyester (which makes up over 60% of all fibres used in clothing), notably from the 1930s, required landfill sites closer to growing, urban populations. With the advent of synthetic materials used for clothing and other product and packaging, a new waste stream has been created that defies re-cycling, re-use and disposal. Archaeological digs at landfills in Birmingham, UK, were first filled with rubbish in the 1950s to 1960s and then covered and 'greened' over (*sic*) to mask their waste treasure, revealed clothes that were made with synthetics. Adult and children's clothes were completely intact, with no degradation even after over 50 years (Evans 2018c).

Obroniwuawu – *From Charity Shops to African Markets*

The recycling of clothes via charity shops and collection points has been a particular Western phenomenon. According to Yvonne Ntiamoah,[9] in Ghana alone, it is

Figure 7.9 Charity Clothes Collection Point, Municipal Swimming Pool Car Park, North
London

Source: Photograph by the author.

estimated to generate an income of £25,000 a day just from UK charity shops, with
imports of 30,000 metric tons a year. As part of the recycling chain for the West,
in return Africa is flooded with cheap accessible clothes, sold in local markets –
Obroniwuawu ('dead Caucasian's clothes') known as the 'Formidable Force' or
Fose, is the local name given to the second-hand clothing industry, one of the main

causes of the collapse of the local textiles industry. The trade runs from the port in Tema through Accra then onto Kumasi (Ghana's largest cities), spreading through all the towns and villages along the route and beyond.

The market women that deal in *Fose* come from generations of traders in the second-hand market, in contrast to the traditional Ghanaian craft-makers of the prized Kente cloth. In the West, the charitable act of donating unwanted clothes to charity shops is regarded as 'good practice' until it is turned into the trade that threatens a whole industry in developing countries. This recycling trade has had a massive impact on industries that have been handed down for generations and has imposed unreasonable restrictions on the fashion and textiles industry in Ghana and other recipient countries. Local producers and designers cannot compete with the prices of the second-hand goods imported into the country and sold in the local markets. If unwanted clothes that are sold in charity shops are limited to the local community where the products are originally donated, they would then maintain their recyclable value, but if sent to communities that have very little income and their livelihood is based on making and trading their indigenous designs, textiles and crafts to each other, then it clearly becomes damaging and unsustainable.

Recycling and technological solutions alone are seemingly not the solution, therefore, faced with the reality that so-called 'biodegradable' material does not actually biodegrade in landfill (starved of oxygen and light); 'technical' or 'smart' textiles which are reliant upon micro-plastics pollute the food and water chain; and as we have seen, the export of second-hand clothes destroys developing country textiles production and markets. A more constructive if small-scale response to the recycle-reuse-repair mantra has also been the resurgence of make-do-and-mend movements. Graziano and Trogal (2017) situate this growing practice as a political/ecological activist phenomenon, encompassing bicycles, electronics, furniture, household improvements, environment (gardens, guerilla growing), as well as clothes and other textiles, with activity taking place not only at home but also at collective and social places such as repair cafes, community tool libraries and online fora. The re-emergence of repair and mending among a generation who had largely lost the skills commonly held by their parents/grandparents is also being stimulated by austerity, greater environmental awareness and a maker movement that is facilitating knowledge and skills exchange. An indication of the now mainstream market for second-hand clothes is the *pre-loved* sections of high street and online retailers, such as Top Shop and Asos Marketplace, alongside their fast fashion offer, and an increase of one and a half times the sales value for second-hand and vintage clothes over the past five years. Fast fashion retailers such as H&M[10] now provide clothes donation points and re-wear lines in their stores, and the extent to which these initiatives will spread and be maintained by the industry will signal their commitment rather than what maybe a time-limited novelty.

Artists exploring the waste-recycle-design nexus include both a long tradition in 'found art' (discarded/waste materials and non-art objects) and also novel transformation of natural and waste material into functional, creative objects (Ehrman 2018). Textiles designer and artist, Kuniko Maeda for instance, works with the ubiquitous brown paper bag as her starting point, which is then treated with persimmon juice to alter its properties, and then laser-cut, producing no wastage. As she observes:

> [W]e seemingly have a feeling of positivity and security to paper recycling without considering actual material value which sometimes causes more consumption. Paper can be more valuable depending on how we communicate with it. I used paper carry bags as my main resource, questioning how to regenerate the value of paper waste, I was motivated to convert disposable and low-quality paper into long-lasting and high quality fashion artworks.
>
> (Maeda, in Evans 2018, 10)

Using this hybrid method, her artwork/garments can be both rigid and flexible, but surprisingly robust, producing headgear, sculpted clothes and interiors that belie their lowly waste origins (Figure 7.10). In contrast, other textiles artists ignore the waste-sustainability issue, continuing to work with petroleum-based materials. Ukrainian artist Asya Kozina for example creates historical and fantastical costumes, but using synthetic paper (resin-based polypropylene/polyethylene/) to build intricate headpieces. In 2019, Dolce and Gabbana bought two of Kozina's creations for a runway show, whilst at the Metropolitan fashion week in Los Angeles, she was commissioned to create a costume inspired by the architecture of her country using the same process. Despite the advances in sustainable materials development, both fast fashion and haute couture have been slow to adopt and adapt to sustainable materials practice, or accept polluter-pays principles in their disposal. Some such as Gucci, owned by French brand-holder Kering, claim to be *Carbon Neutral* in their marketing, but this is a case of greenwashing with credits purchased to offset emissions arising from the company's production, a practice recently banned by the Advertising Standards Authority.

Fashion is therefore 'a clear demonstration of place-specific comparative advantage and specialisation, intensely linked to place in product' (Williams and Currid-Halkett 2011, 3043), and waste arising from clothing is, on the one hand, extremely

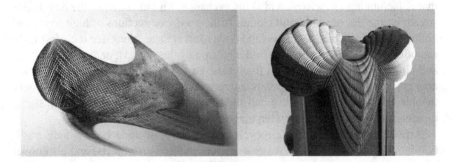

Figure 7.10 Kuniko Maeda Fashioned From Paper Bags
Source: Photograph by the author.

located – situated, culturally context-specific – but, on the other hand, it is also *dis-located*, with an economic, ecological/ethical footprint stretching across space and time. Place in particular makes a difference to both how fashion production and identity are formed and projected, and also how fashion-related waste gets generated and in what context and scale. Fashion – and its material waste – is therefore an embodied cultural product, and awareness of its content, disposal, re-use and recycling challenges and possibilities can make a huge difference to how it is perceived and managed in the future.

*

Whilst fashion garments may be a second-skin to the wearers, their design influences have come from art and increasingly from the street – and not surprisingly, street art. Fashion brands have taken a significant interest in street artists within the past few years, but this cross-over between designer and artist is not new, for example, Keith Harring's images have been an influence to designers (and on display at Tate gallery's Keith Harring exhibition in 2019 was a Vivienne Westwood and Malcolm McLaren knitted jumper and skirt ensemble). In 2017, Vogue, the *fashionistas* bible, carried an article in defence of street art[11] lauding the political power of the acceptable face of graffiti, a sign that fashion and street art at least, had common cultural influence and importance. The next chapter is therefore dedicated to the cultural phenomenon of graffiti, and its more acceptable incarnation, street art, representing the ultimate public gallery space.

Notes

1 https://fashionunited.com/landing/fashionweeks-around-the-world-list
2 www.thembsgroup.co.uk/internal/streetwear_and_couture__blurring_lines_in_the_ fashion_industry
3 Sherwood, H. (2023) 'From rebel to fame: public gets access to David Bowie archive', *The Guardian*, 23 February, 11.
4 https://beyondbanglatown.org.uk/globe/working-lives-clothing-to-catering/
5 www.artscouncil.org.uk/lets-create/delivery-plan-2021-2024/strengthening-our-place-based-approach-and-supporting-levelling
6 www.centreforcities.org/reader/move-public-sector-jobs-london/summary
7 https://wrap.org.uk/resources/guide/circularity-fashion-and-textiles-businesses
8 Whilst textiles are only one of many "cultural products" – others include product design, E-/or i-Waste and an impending environmental catastrophe from energy-hungry computer servers and data centres as well as rare/toxic metals in mobile phones and other electronic goods – the sheer volume and turnover of clothing and synthetic components present a particular social and cultural challenge.
9 https://portfolio.arts.ac.uk/project/176541-board-of-fashion-ghana-new-generation-of-designers-embracing-circular-and-sustainable-fashion/
10 https://www2.hm.com/en_gb/sustainability-at-hm/our-work/close-the-loop.html
11 www.vogue.co.uk/article/in-defence-of-street-art

8   Graffiti and Street Art

Graffiti and its more recent categorisation, Street Art, represent a particular cultural space and practice, with a millennia-old presence and contemporary interpretation. Indeed, the term street art as opposed to vandalistic graffiti indicates both a valorisation and cultural shift in public art itself. Areas that are the subject of intensive and expansive graffiti and street art could be said to represent museums, or at least galleries 'without walls', creating cultural markers on surfaces that were not intended for such use. As such, graffiti and street art production sit uncomfortably between architecture (surface), advertising (the artist/group) and the culture (identity) of a place. This identification can be highly localised, or like much mainstream culture, highly globalised – or at least, seemingly familiar.

Graffiti, and even the more refined street art, tends to cluster and repeat itself. This is not surprising given the local conditions usually required – available 'wall space' and a public viewing location (without the potential for an audience, graffiti cannot exist other than as some personal record, e.g. on a prison wall) and the scope for working without the risk of being caught red handed and/or prosecuted. But the presence of graffiti attracts further graffiti, often over-writing or extending, whether competitive or collegiate in motivation. Graffiti as a visual manifestation of vandalism has also been associated with urban decline, a heightened risk of crime and threat to community safety at a neighbourhood level. Broken window theory for example suggests that as more dereliction takes hold on buildings, this generates a spiral of decline as neglect leads to more deterioration of the built environment, including unchecked graffiti (Wilson and Kelling 1982). However, the transformation of graffiti to street art as a social and visual practice challenges these assumptions, with street art as much a visitor attraction and celebration rather than a symbol of decline, and no longer associated with vandalism and loss of control. The contract (sic) obviously assumes that property owners and occupants, as well as local agencies – police, local authorities – at least tolerate street art and in some cases encourage it, as part of placemaking and regeneration strategies. In practice, however, this is a meanwhile situation in the period between building redevelopment and re-use, where occupiers are short-term (including artists). In areas of long-term regeneration, this can amount to several years, whilst, in other areas, street art is normalised and part of an accepted practice and presence, where there is stability in the property market, at least in the medium-term. Here

DOI: 10.4324/9781003216537-8

graffiti can be seen as an architectural element in building façades or equivalent to advertising billboards and signage. Longer term however, gentrification can eventually 'clean up' an area or particular buildings and estates, as rents rise and private corporate interests trump area-based identities. There are parallels therefore with the treatment of street art, and the occupation of redundant former industrial spaces by artists – to use the apt quote from Montreal graffiti which was reproduced on the cover of the Toronto Arts Council's corporate report: 'Artists are the storm-troopers of gentrification' (TAC 1988; Evans 2001, 172). Today, street artist CSRK has been fusing well-known film characters with the streets of Montreal. From Stormtroopers holding cans of poutines and construction signs to Han Solo's blaster gun being replaced with French fries, cheese curds and gravy, the force has become strong on certain walls around town – a blast perhaps of post-modern irony, or just opportunism.

Graffiti and Street Art have received a variety of research treatments, from artist, subcultural, ethnographic and crimogenic perspectives, which are reflected in the literature on individual graffiti and street artists, gangs and genres, with a growth in art monographs and coffee-table style pictorials. As Irvine observes: 'street art is . . . defined more by real-time practice than by any sense of unified theory, movement, or message' (Irvine 2012, 235), and there has been a distinct lack of empirical and theoretical critique, including the role of place and reception (Nitzsche 2020). The productionist dialectic between graffiti as vandalism and as street art (including in galleries and auction houses) has not considered the wider aspects of the role and place of graffiti and street art in the city; responses from city authorities, local communities, visitors and property owners; and how different places and city cultures receive and react in different ways, including graffiti and street art now used in placemaking and destination strategies. This includes the growth of graffiti and street art commissioning agencies and organisations (e.g. *Book an Artist*), often established by former graffiti and street artists employed by clients such as retail and advertising firms (Borghini et al. 2010), local authorities, and legitimised spaces and walls for safe experimentation. The cultural content of graffiti and street art also reflects local conditions and contexts, whether protest, political, territorial, vernacular (e.g. local events, history) or playful in nature. Graffiti and Street Art now occupy an enduring place in the image of the city and our urban heritage, and not surprisingly increasingly feature in city tours and trails:

The tour bus picked us up outside of the designer-hotel in Manhattan. Commuting office and shop workers, tourists, police and road diggers mingled in the chaos of downtown traffic. Across the Williamsburg Bridge we stopped to pick up our tour guide for the day, Angel Rodriguez, from an insalubrious building covered in layers of posters, graffiti and grit. He was a Latino musician, a salsa drummer from the Bronx who proceeded to give the tour group the background to the area – 'Bronx is burning' (arson attacks on tenement blocks by landlords), old jazz and dance club haunts, Fort Apache – the movie – the now rebuilt district police station self-styled to defend itself against the 'natives', (i.e. black/Hispanics), graffiti art of local rap stars (Figure 8.1), the massive American Mint building covering four blocks, where two thirds of all US dollar notes were once printed, now housing two community schools, artist studios and employment schemes; the local penitentiary with twelve year olds kept in shackles – before arriving at our destination, The Point. Here the graffiti boys' operation base – once the crew that covered the New York subway trains and led to the mayor's zero tolerance regime – have now gone 'legit', working for advertising firms and department stores in Manhattan on large-scale shop displays and billboard art.

(Evans 2007, 35)

As street artist Matteo suggests:

In effect, we have something in common with advertising . . . a deep link. Not only the format: large posters and billboards . . . also the place: on walls, like advertising. The invasive effect is the same, the attack is the same.

(Borghini et al. 2010, 125)

Graffiti has thus come a long way from its modern roots in the 1960s, commonly attributed to tagging/name signatures in ethnic ghettoes of New York, Philadelphia

Figure 8.1 Big Pun 'Memorial', The Bronx, New York

Source: Photograph by the author.

and Los Angeles, although parallel outpourings were evident in other cities such as London (Dawson 2021), including protest and wall-poster art, but this was hardly a new phenomenon, since *graffito* were as universal 400 years ago as they are today, and just as disreputable (Jones 2021). What this has signalled, however, is the beginning of a commodification of this activity, with a widening of graffiti into other cultural forms such as Music (e.g. rap, with graffiti one of the four 'core elements of hip hop', Nitzsche 2020), Fashion (e.g. street artist *Mau Mau* graduated from T-shirt design to graffiti art), Film (e.g. animation, pop videos), Visual/Pop Art, Advertising, and Architecture/Urban design, which collectively has extended its shelf life. Early forays of artists who first worked on the street and in street-style into art galleries and public venues had been less successful, despite the rapid valorisation and curation of particular artists work in recent years, notably the late artists Jean-Michel Basquiat and Keith Haring in the USA, and in the UK, Banksy, whose distinctive stencil murals have fetched over $500,000 (often to US buyers and celebrities). This reached a nadir in 2014 – from Banksy's street-cred perspective – in an unauthorised retrospective of 1970 'works' by the international auction house Sotheby's in London. Here, however, the work has been first validated in situ (a fundamental element in its value and authenticity), and then removed, much like a historic mural, into a private collection.

Graffiti has therefore largely resisted (art) museumification and thrives primarily in a museum-without-walls – but very much on a city's walls. The varying treatment and cultural significance of graffiti were also observed by Walter Benjamin, in his essay on the walls of 1920s Marseilles:

> Admirable, the discipline to which they are subject, in this city. The better ones, in the center, wear livery and are in the pay of the ruling class. They are covered with gaudy patterns and have sold their whole length many hundreds of times to the latest brand of aperitif, to department stores, to the 'Chocolat Menier', or Dolores del Rio. In the poorer quarters they are politically mobilized and post their spacious red letters as the forerunners of red guards in front of dockyards and arsenals.
>
> (1999, 135)

Indeed, it was the appropriation of the post-industrial city's walls by what are perceived as undemocratic and unwanted advertising images that has provided the political impetus for contemporary graffiti artists such as Banksy with graffiti 'principally an anti-art movement, making art form out of vandalism' (Armstrong 2019, 15). As Banksy himself proclaims:

> Twisted little people go out everyday and deface this great city. Leaving their idiotic little scribblings, invading communities and making people feel dirty and used. They just take, take, take and they don't put anything back. They're mean and selfish and they make the world an ugly place to be. We call them advertising agencies and town planners.
>
> (2002, n.p.)

Cronin goes as far to suggest that outdoor advertising and graffiti should be studied together, in terms of their ubiquity and visual impact (2008). As Borghini et al. also observe:

> Street art is a product that embodies its own advertising . . . a countercultural response to commercially or statist-induced alienation, street art as a populist aesthetic, a consumerist critique, and urban development project. Street art espouses a vision of space reappropriated as place, where commercially noisy or entirely silent streets are reclaimed by artists for their 'owners'.
>
> (2010, 113)

Zieleniec, in analysing graffiti through the prism of Lefebvre's *right to write the city*, optimistically asserts that: 'graffiti makes space social and public, through the promotion of use values and meaningful acts of colonisation and inhabitation versus the homogenising practices of planning, design, commerce, and their overarching concern with surveillance, order and security' (2016, 2).

Graffiti as Vandalism

Until recently, official responses to graffiti have placed it squarely in the criminal 'vandalism' sphere and early commentators fuelled this view: 'graffiti disrespects private property and official notions of order and aesthetics' (Lachmann 1995, 100 on Ferrell 1993). Early responses to the graffiti 'epidemic' in New York and Los Angeles saw criminal sentences increase and special tasks forces established, claiming that the order of the landscape had been disrupted, and clean-up costs were rising: more than $50 million a year in both cities by the late-1980s. In the UK, clean-up of graffiti is estimated to cost £1 million per year and in Chicago alone, $6 million (see graffitihurts.org). English Heritage estimates that 70,000 heritage buildings and monuments are vandalised and defaced by graffiti, with porous stonework, bricks and lime especially vulnerable and not able to take anti-graffiti protective paints since they are not suitable to these types of heritage material.[1] Elsewhere in the UK, Network Rail spends £5 million and London Transport £10 million a year on graffiti clean-up.

In 1960s/1970s New York, gang graffiti-ists were also enabled by the subway system that took their tags across all of the city's boroughs and away from their local territories, with large *pieces* covering whole carriages (Transport Police advise operators to take trains out of commission quickly to reduce this incentive). Whilst the New York subway was successfully cleaned-up, transport still remains a key site for graffiti – attractive for its wide availability and high audience potential. For visitors to many cities, whether by road or rail, the first visual sign they will see is graffiti and tags along motorway walls and bridges, and on the interstitial approaches to railway stations (e.g. Maastricht). Despite the advent of CCTV and other surveillance, stations and bus stops receive both unwanted as well as commissioned artworks and tagging, for example in Stockholm *Art on the Underground* (Figure 8.2) and in London an on-going programme of *Art Posters* and *Poetry on*

Figure 8.2 Art on the Underground, Stockholm and Graffiti on Amsterdam Bus Shelter
Source: Photographs by the author.

the Underground commissions. Graffiti and Street arts have therefore been faced with a dual onslaught from different dominant forces – police/city politicians and art curators/galleries – to remove or control its practice and impact. However, despite this, or perhaps spurred on by this marginalisation, counter-hegemonic discourses have emerged which in some senses have kept graffiti alive as both a cultural concept and a practice that is now evident in many forms internationally – that is, graffiti is now a global cultural phenomenon. Lachmann's observation in 1988 is therefore still valid today: 'Graffiti in some forms can challenge hegemony by drawing on particular experiences and customs of their communities, ethnic groups and age cohorts, thereby demonstrating that social life can be constructed in ways different from the dominant conceptions of reality' (1998, 231–232).

This challenge is evident in graffiti artists' response to the art market itself, in the case of Banksy's *mockumentary* film *Exit Through the Gift Shop* (2010). Here, a fictional filmmaker pursues the underground art scene in Los Angeles, New York, London and Paris, assuming the role of self-styled street artist hyping his avant-garde 'show' in LA and creating an art world/underground buzz for the lucky few who could take part:

> Banksy thus pokes fun at the contemporary art world and its hunt to unearth and exploit underground art scenes. The willingness to validate recycled art and popular cultural symbols, which are rendered empty if not meaningless, is revealed as undiscerning and opportunistic . . . whereby social critique is downplayed in the pursuit of print, poster, and postcard sales.
>
> (Birdsall 2013, 116)

The mobile value of his own street art has however fuelled a destructive market (another form of vandalism) which sees publicly displayed works cut from walls to disappear then reappear via auction. These sites continue to attract viewers and subsequent graffiti responses, and place branding through graffiti can therefore persist (including in memory) despite the absence of the artwork itself following its removal, increasingly for financial gain (although not for the author/originator). Leader recounts the story of the theft of the Mona Lisa from The Louvre in 1911 – after the painting was stolen, thousands of people flocked to see where it had once been on display. Many of them had never seen the painting in the first place (Leader 2002), and this phenomenon is repeated each time a Banksy is discovered and 'lost' today (Hansen and Flynn 2014), with this cycle recurring each time a Banksy appears on a private or municipal wall – local excitement about putting their place on the map; followed by national media attention and a flow of visitors; and inevitably its removal/theft and pilgrimage to the site-of-absence. Meanwhile, commercial gallery owners vie for acquisition with offers to an unsuspecting garage owner (a local steelworker), in the case of a recent Banksy which appeared in the Taibach neighbourhood of Port Talbot, Wales. The piece named *Season's Greetings* stencilled on to the corner of a breeze-block garage close to the local steelworks, stood for one year, before being bought from the garage owner by an Essex art dealer who temporarily relocated the 4.5 tonnes of a wall section to a former police station nearby. Two years later, the piece was removed from Wales altogether with a candle-lit vigil for its departure featuring a local poet, *Port Talbot's Got a Banksy*, children singing *Little Snowflake* – the tune Banksy's people dubbed over video footage of the mural – and speeches imploring the piece stays in the town. According to one of the organisers (with no sense of irony): 'we can claim to be the street art capital of Wales. We may not have the funding to build a big art gallery, but we have the power to turn our streets into a gallery' (Morris 2022).

Mixed Messages

Today, this duality, vandalism or art, continues, reflected in prohibitory sentences – in the UK up to ten years imprisonment where criminal damage by an adult (18 years) exceeds £5,000, and detention/training order of up to two years for 12- to 17-year-olds. For minor offences, sentences are much lower and fixed penalty charges can also be issued (up to £100) without court proceedings, so there is some discretion over the response if found guilty/*caught in the act*. At the same time, art museums and galleries engage with graffiti as an international art form. For example, in 2008, Tate Modern's *Street Art* commission and exhibition brought together six internationally acclaimed artists whose work is linked to the urban environment. Sponsored by the Japanese car firm Nissan, this was the first major public display of Street Art in London. In order to give it artistic validation, 'good' street art in this case was distinguished from the more low-brow graffiti and tagging, thus seeking to 'insert graffiti into its proper place and rob it of its denaturalising power' (Creswell 1996, 55; Creswell 1992). The link with sponsor Nissan was also significant, since the Qashqai car it launched the year before utilised striking street art in its adverts. More recently graffiti has also cemented its place in the commercial art gallery world with Saatchi's three-storey gallery hosting the largest ever exhibition of street art in the UK, *Beyond the Streets London,* also sponsored this time by adidas sportswear, and succeeding *Beyond the Streets* exhibitions in Los Angeles and New York.

This latest celebration of over 100 street artists work: 'examines the fundamental human need for public self-expression, highlighting artists with roots in graffiti and street art whose work has evolved into highly disciplined studio practices, alongside important cultural figures inspired by this art scene'.[2] This also provides a clue to the current ambiguous place and relationship between graffiti/art and the city. In one sense, this reflects the consumption and visual culture prevalent in the contemporary city environment – the merger of commerce and culture in highly visualised form. As Chang maintains: 'saturated by images, the contemporary city has been theorised as a site legislated by the eyes' (2013, 216), whilst street art today in Irvine's view 'is a paradigm of hybridity in global visual culture' (Irvine 2012, 235).

Lombard goes further in response to the question, has the governance of graffiti changed since its more reactive origins? She uses the concept of governmentality (2013), or what Gordon coined, the *conduct of conduct* (1991), to analyse how graffiti is currently controlled, arguing that whilst there appears to be a softening of policy and responses towards graffiti, this does not mean that there is less governance, but that this marks a greater acceptance of graffiti due to the effects of a neo-liberal form of governance. Chang also notes the emergence of the counterveilling terms: post-graffiti and neo-graffiti:

> signalling some kind of qualitative and stylistic shift in modes of inscribing the city. Encompassing multiple forms or urban inscription like murals, postering and sculpture that move beyond written text . . . (which) mark the spectacular nature of urban space.
>
> (2013, 217)

Here, she critiques the work of artist Blu, who painstakingly paints and photographs over existing graffiti (representing a single film frame) then turns these into remarkable street life animations (see www.blublu.org). This is one example of graffiti being transformed into moving image whilst drawing and building on its everyday street art nature. This also represents an important cultural practice of capturing as well as creating graffiti art in a different non-ephemeral form – important with so much street art being time-limited and subject to clean-up, defacing and deterioration due to the weather, etc. Archives of graffiti art also seek to document this work alongside publications and films, in a variety of media (DeNotto 2014).

A governmental response to the 'demand' for graffiti by young people is also seen in various schemes which seek to offer a safe (from prosecution) opportunity for budding Banksy's to practice their art with impunity. For example in Wales, *The Heritage Graffiti Project* helps young offenders 'learn valuable lessons from their heritage'[3] by introducing them to archaeological artefacts and explaining what they mean to the people who used them (e.g. Roman soldiers, miners, canal boaters). A mural was created, *Our Wales*, by the young people depicting their interpretation and experience that was documented and opened to the public. In the DPM Park in Dundee, Scotland, the longest legal graffiti wall in the UK at 110 metres is open for all to use at any time, and the council-run project holds workshops for local kids. How far these participants subsequently refrain (or not) from illegal graffiti activity is not, however, measured. Glasgow is the latest Scottish city to celebrate and promote street art with an annual *Yardworks* festival held around the SWG3 arts venue in a former galvanisers yard near the River Clyde. Here, the city council is also exploring setting up legal walls where graffiti artists can develop their work without fear of arrest. This in a council which spends double its nearest rival (Hackney, east London – see in the following) on graffiti removal. From the perspective of Glasgow street artist, James Klinge: 'people can think of a gallery as a really intimidating place to go into, but anyone who can walk down the street and see this mural in its progress – that's their art gallery' (Brooks 2022, 7).

Attitudes of city residents towards graffiti and street art are also changing and ambiguous. These, not surprisingly, also vary within cities, with some neighbourhoods, sites and buildings treated differently in terms of surveillance, prosecution, protection – and celebration. For example, in the Colombian capital, Bogota, following the death of a young street artist shot by police in 2011, a new tolerance of street art has emerged. The city mayor issued a decree to promote the practice of graffiti as a form of artistic and cultural expression, whilst at the same time defining surfaces that are off limits, including monuments and public buildings. City grants are available for selected artists with two, three and even seven-storey walls provided along the main thoroughfares as their canvases. The result is colourful displays with political and social messages. Everyday graffiti has also spread under this liberal regime including on buildings prohibited from writing. This indicates the difficulty of controlling graffiti in this way without rules being observed, and the appeal to marking untouched surfaces and public spaces. Lisbon is another city that has embraced street art, as it tries to move out of austerity and recession (GAU 2014). Large-scale pieces and murals adorn public buildings and industrial districts (Figure 8.3), which range from artistic to protest images and messages.

Figure 8.3 Street Art, Lisbon
Source: Photograph by the author.

Here, local artists, such as Vhils (Alexandre Farto), are celebrated and combine the aesthetics of vandalism with social comment. Rather than spray can and stencil, Vhils carves into the render of the walls of buildings using electric drills to produce large-scale portraits, often of local community members, and also works on utility installations. As a sign of his acceptance into the city art establishment, his inaugural exhibition was held at the opening of a new art museum based in a converted electricity station *Museu Da Electricidade* (Figure 8.4). Lisbon University also hosted the first international conference on Street Art and Urban Creativity (www. urbancreativity.org) with an extensive programme of papers from academics, curators and doctoral students.

Elsewhere, commissioning of young artists to adorn corporate buildings presents an alternative to the typical public art installation. In Frankfurt, the European Central Bank HQ, a 45-storey building then under construction, was surrounded by a high protective fence. A local social worker approached the bank that agreed to allow him and the 'troubled' young children he works with, to spray paint a wooden fence that they erected around the site (costing 10,000 Euros). The graffiti depicts caricatures of the ECB President and the then Chancellor Merkel (60% of the works reflected the Eurozone crisis), and fighting cocks that were

Figure 8.4 Electricity Museum and Storage Tank Graffiti (Vhils *Dissection* Exhibition), Lisbon

Source: Photograph by the author.

displayed within the building when it opened. Several of the graffiti artworks have been purchased via the *Under Art Construction* programme, ironically, by bankers, although remaining works are not, apparently, for sale. The city mayor has called on other construction sites to emulate this project. Other *meanwhile* sites are also the subject of sanctioned graffiti, since these can on the one hand animate otherwise ugly hoardings and also prevent/dissuade opportunistic graffiti, as well as divert attention from permanent structures. For example, in Madrid and in Amsterdam, where the former Royal Dutch Shell European HQ building awaits redevelopment with temporary occupation by dance event organisers (Figure 8.5). This northern part of the city known as Amsterdam-Noord also represents a new creative quarter, served by frequent free ferries from behind the main station, where a cluster of digital media workshops and arts and entertainment venues has replaced this industrial complex and working-class district (see Chapter 5). In this case, graffiti art might signify transition, fun and creativity – rather than degradation and social unrest, as it may have done in the past. In many situations, it signals a start to a gentrification process whether at a meanwhile stage, as part of resistance and presaging new developments, or even celebrating the 'new' or newly conceived place itself.

Figure 8.5 Graffiti Art on Base of Former Royal Dutch Shell HQ, Amsterdam-Noord

Source: Photograph by the author.

The tensions between crime-art and control-tolerate in practice are therefore played out in a continuum along which city authorities, the public, and graffiti and street artists move, as taste, opinion (including local and national media), city branding and development shift over time. This can represent a hardening as well as a softening and instrumental use of street art, as we have seen, increasingly used in city branding and placemaking efforts and strategies. The public is, of course, no longer homogenous as major cosmopolitan cities and historic towns mix tourists and a range of business, education and leisure visitors with residents and commuting workers from many countries with differing aesthetic and moral positions. The perspective of, say, an overseas tourist to street art/graffiti, may be one of attraction, branded image, signifier of a cool place – or one of fear, decline and poor aesthetic value/appeal. To a resident, the same images may form part of their everyday experience, represent local identity (theirs or others – good, bad or indifferent), or even align with the visitor's view. These thus become heterotopian spaces in the Foucauldian sense, representing a particular place, but one experienced over, and in, space-time (Massey 1999; Unwin 2000).

It is more likely, however, that the local resident will engage in a deeper, knowing way, depending on the length of time the graffiti has been there, where it is placed (i.e. on what type of building/structure) and its meaning to them, if any. Graffiti and street art are certainly increasingly identified with a sense of place than was the case before – aside from the previous tags and territorial/gang variety which are more likely to be cleaned up by city authorities. The attraction of place to graffiti artists is reciprocal. Cities in flux such as post-reunification Berlin were perceived as a 'graffiti Mecca of the urban art world . . . the most "bombed" city on Europe' (Trice, in Arms 2011). Here its acceptance/condonement was associated with Berlin's designation as UNESCO City of Design and as a growing cultural tourist destination, which is in part fuelled by this urban image of street creativity. This includes international artists, including graffiti artists whose work moves from street to gallery to street and of course via social media. In cities such as Sydney, the creative class discourse (Florida 2005) and policies towards public art have also provided 'opportunities to resignify graffiti as productive creative practice' (McAuliffe 2012, 189).

In cities, and therefore in placemaking efforts, urban space is lived through its associated images and symbols (Lefebvre 1991) together with other senses, and which together stress the essential experiential nature of the relationship with our everyday environment, and our identification with discrete places and spaces. In this sense, we do not 'use' space or our urban environment as consumers (as with branded products), but we experience it individually, productively (i.e. in/through work) and collectively, albeit with diminishing influence over the (re)construction of the public spaces we inhabit – including the presence of graffiti. What is becoming evident therefore is that graffiti and street art have become established visual forms which cities both adopt and project as part of their identity and their destination marketing mix. Street art is therefore now an emerging strategy for placemaking and branding particular areas, which takes it out of its crimogenic roots (Ross 2016), as these examples illustrate.

Hackney Wick, London

Hackney Wick in east London hosts a high concentration of practicing artists who work from studios, among temporary gallery spaces and a cluster of industrial buildings and canalside infrastructure. This concentration has been accelerated as studios have been demolished to make way for the adjoining Olympic facilities and park and new housing in the post-event phase of development, and as the cost of workspace has increased in other parts of east London (priced out by commercial housing and hi-tech markets as discussed in Chapter 6). This landscape and industrial canvas have provided an opportunity for graffiti artists to create large-scale works (Figure 8.6) and also to express their displeasure at the gentrification of this neighbourhood, which may also lead to their eventual displacement (Figure 8.7).

The area hosts an annual arts festival, Hackney WICKed (see Chapter 9), and is promoted as a visitor destination by the local authority, the Canals and Rivers Trust, and the Olympic Park authority, who see their mission as 'stitching together' these divided neighbourhoods and communities, in the words of the local development corporation: 'through design quality to create a unique and inspiring place for events, leisure, sport and culture, a hub for enterprise and innovation, and diverse sustainable communities' (LLDC 2014, 5). The agencies that control and legitimise this newly promoted district have also commissioned graffiti artists to adorn local buildings as part of a curated[4] initiative which engaged international (as opposed to local) artists – for example the Canal Project[5] (Figure 8.6).

This project was however met with outrage by local graffiti and other artists and residents, following what had already been a graffiti clean-up of the area prior to the 2012 Olympics. To add insult to injury, several artists from Brazil, the Netherlands, Sweden and Italy were funded to produce artworks on these same buildings. In the words of the funding agency: 'we unashamedly wanted to showcase the best international artists and transform this part of the canal into a destination for street art – I hope people will come on boat tours to see the work' (in Evans 2016d, 178). But as local graffiti artist *Sweet Toof* says: 'with the commercialisation of street art it's becoming pay-as-you-go wall – every surface sold off to the highest bidder' (Wainwright 2013). Placemaking which adopts graffiti as a distinction and vernacular expression will therefore need to develop strategies which treat such work as part of the area's built heritage and community culture,[6] as Smith observes: 'It's as if the street art has been given the responsibility of preserving the Wick's soul as it's squeezed on all sides by colossal tectonic pressures of redevelopment' (Introduction, in Lewisohn 2013, 5).

Shoreditch

The image of this neighbourhood combines post-industrial use of workshops and factories, with small crafts and retail outlets, social and warehouse loft apartments and extensive graffiti art on this historic mix of buildings and walls. This has offered an effective graffiti street laboratory within which aspiring artists such as Banksy and Stik first experimented (Figure 8.8). As an indication of the importance

Figure 8.6 Canal Project Commissioned Graffiti Art, Hackney Wick
Source: Photograph by the author.

Figure 8.7 Graffiti Art, Hackney Wick
Source: Photograph by the author.

Figure 8.8 Shoreditch Graffiti: Stik, Banksy and Dscreet
Source: Photographs by the author.

of street art, several companies provide guided Shoreditch Street Art tours, with online galleries and listing of artist/artwork profiles – *In Shoreditch art is an open-air affair. From huge murals on buildings* (see 'Bees' in Figure 8.9) *to tiny stickers you'll spot everywhere, the streets are fair game. Who knows, you might even spot a new Banksy* (Shoreditch Urban Walkabout 2014).[7] As in New York, specialist galleries and agencies also provide commissioning services for clients wishing to hire graffiti and street artists for temporary or permanent work – such as *Graffiti Life* and *Graffiti Kings*. The time of the professional graffiti artist and intermediary has therefore come, whilst artists have long taken graffiti into their art works, for example, resident performance artists Gilbert and George used graffiti images from the area in the 1970s in their work *The Dirty Words Pictures* that juxtaposes graffiti swear words and slogans with disturbing images of urban life and the bleak presence of the artists themselves.

Figure 8.9 Bees Graffiti, Shoreditch
Source: Photograph by the author.

This inner urban district has evolved from a working (class) workshop area servicing the City and its low-income residents to a techno-creative habitus that 'presents itself as a loose association of the *digerati* for whom "Shoreditch is a state of mind"' (Foord 2013, 57), and within which established graffiti artists mingle and make their street art. With the off-beat night-life and café culture supported by this local ecosystem, artists are also guaranteed a growing visitor audience, however, as Foord also observes: 'The gulf between the aspirations of the *digerati* and those of local communities who are confronted daily by the impacts of recession are unlikely to lessen' (ibid., 59). Graffiti artists increasingly have an ambivalent place in this situation and have in effect joined the existing producers of space.

If You Can't Beat Them, Join Them

Graffiti and Street art thus have a complex and ambiguous place in the city. Clearly a duality now exists between 'high' street art and (un)popular graffiti. Technically an illegal activity unless fully commissioned and authorised by property owners and other stakeholders (notably transport and local authorities), condonement of street art is evident in cities and areas of a city where either control has diminished or a general laissez faire situation exists. This is currently evident in cities where economic decline and socio-political fragmentation has reduced the power and resources for clean-up or enforcement, for example Athens (see Avramidis 2012). Here, the vacuum this has created is also fuelled by political response/resistance to the governance deficit and economic impacts, (e.g. unemployment, debt, cuts in services). Other cities equally affected by the severe economic recession have adopted a more creative approach, such as Bogota and Lisbon, as discussed earlier.

Attitudes of local police are also variable and their stance on graffiti and street art can be determined by a number of factors. In the USA, research on a mid-Atlantic police department found that the race of police officer and the shift – day or night – effected the attitude towards graffiti crime, and therefore towards perpetrators and enforcement (Ross and Wright 2014). Safer Neighbourhood Teams in London also follow a priority crime regime, as a form of resource efficiency and policy/political targeting (e.g. burglary, mugging), leaving graffiti deprioritised unless literally caught in the act, or in response to complaints. This contrasts with the British Transport Police who operate a zero-tolerance regime, recording and attributing graffiti to identify subsequent offences and provide evidence to support prosecution of what they term 'serious vandals'.[8] Police Officers in Scotland also collect *tags* into a database which is then used to pinpoint where a particular image is being used so that officers can narrow their hunt, which has led to several arrests.

However, in areas undergoing transformation or interstitial post-industrial zones, where landowners are distant or unconcerned (and property values are not threatened), graffiti and street art flourishes, as in Hackney Wick and Shoreditch. More cosmopolitan neighbourhoods such as Beyoglu in Istanbul also provide a concentration of street art in a more liberal, if contested, territory than permitted elsewhere in the city (Erdogan 2014). In other cities, areas such as Amsterdam's main university district are subject to extensive graffiti, indicating a combination of tolerance, complacency and placemaking by its mostly student residents. This is also evident in more protest-oriented university zones in cities such as Athens, but this also extends to areas around government buildings and conflict zones, including sites where the death of protesters has occurred (Avramidis 2012). Elsewhere, street art is seen in commercially driven commissioning, installations and contemporary art interventions in downtown, retail and in other locations undergoing gentrification (e.g. Dumbo, Brooklyn New York), particularly in temporary sites. Graffiti still persists, however, as a dominant image in derelict sites and in 'accessible' (*sic*) transport facilities and is still associated in this case with decline and redundancy. In other areas, street art reflects a creative quarterisation of a neighbourhood and effectively helps to add value to its image and distinctive brand. This is not limited to larger cities and urban centres however, in central Portugal the two-week long Covilho Urban Art Festival was first held in 2011 in this rural university town overlooked by the Serra da Estrela mountains. Here, the *WOOL on Residence* brings together national graffiti artists and local artists to produce murals across the town, including references to the area's sheep-rearing roots and wool textile design and manufacturing (see Burel factory, Chapter 7).

Thus street art has, on the one hand, joined the canon of contemporary art, and the art market – if treated with caution by graffiti artists themselves (Brighenti 2010) – and been appropriated in commercial advertising and media, and on the other hand graffiti, in its basic form, continues to inhabit the everyday city environment as a low level 'noise' and nuisance for many, as well as an almost endless canvas for its producers and hosts.

An indication of graffiti and street art's arrival and enviable status is provided by celebrated contemporary British artist, Grayson Perry, on the launch of the *Art Everywhere* scheme which placed selected artwork images on over 30,000 billboard and poster sites across the UK: 'given that street art was everywhere these days, it was nice to put gallery art on the streets' (Brown 2014, 11) – or if you can't beat them, join them. Emulating this initiative, for a month in 2023 the work of eleven artists also appeared on thousands of billboards and digital screens across the UK (sponsored by advertising site companies), which aims to 'create a new kind of cultural institution, challenging traditional models of viewing and thinking about art. A different kind of national gallery: free and accessible, without walls . . . out in the "wild" not in a gallery' (O'Callaghan 2023, 2). Street art has of course been doing this for decades, but in a non-curated, unself-conscious manner.

*

Graffiti is characterised not only as a low form of vandalism but also as an art form and antidote to the commercialisation of urban space. Its activist roots and practice are also apparent, from explicit political statements to local protest and commentary – very much a lived experience that has moved beyond a *hit and run* style, with graffiti artists more directly engaged with cultural development and democratic movements. Not surprisingly, street art increasingly features in socially engaged arts initiatives and in more participatory rather than the individualistic (and anonymous) practice with which it has long been associated. The final chapter explores the practice and place of socially engaged cultural space development and collaborations with host communities in and around cultural and heritage venues, and with practicing artists – including those working *on the street.*

Notes

1 www.researchgate.net/publication/260625858_Performance_and_durability_of_a_new_anti-graffiti_system_for_cultural_heritage_-_The_EC_Project_GRAFFITAGE
2 www.saatchigallery.com/exhibition/beyond_the_streets_london
3 Heritage Graffiti Project http://cadw.wales.gov.uk/learning/communityarchaeology/heritage-graffitiproject/?lang=en. Accessed 5 April 2014.
4 The curator of the Hackney Wick Canal Project was Cedar Lewisohn, who also curated the 2008 Tate Modern Graffiti Exhibition.
5 www.canalrivertrust.org.uk/art-and-the-canal-and-river-trust/the-canals-project-street-art-on-thewaterways. Accessed 3 August 2014.
6 Merrill goes further, suggesting that graffiti and street art should be perceived as an example of 'alternative heritage' whose authenticity might only be assured by avoiding the application of official heritage frameworks (2014).
7 https://mappinglondon.co.uk/wp-content/uploads/2014/09/UW-shoreditch_v3.pdf
8 www.btp.police.uk/advice_and_info/how_we_tackle_crime/graffiti.aspx. Accessed 3 August 2014.

9 Socially Engaged Practice and Cultural Mapping

The cultural spaces considered in the previous chapters have encompassed cultural amenities and *assets* – from sites of heritage which are the subject of selection, and those associated with intangible and community traditions – to places for arts and events engagement and consumption, and spaces of production: digital, fashion and street art, that in different but increasingly convergent ways have come to represent urban 'visual' cultural space. The model of socially produced space operates at several levels and as Lefebvre maintained, these spatial states are experienced simultaneously and existentially – and are not mutually exclusive. Of course, social and technological change since 1960s Paris has transformed production and consumption reality and practice – notably mass tourism, migration and the globalisation of late-capitalism that has seen the further commodification and privatisation of the public realm and the fragmentation of everyday patterns of life – work, recreation and lifestyle – all influencing cultural production and consumption possibilities. Notions of rights to the city and of cultural rights in particular, have inevitably shifted as a consequence, arguably democratisation in some areas, but experienced unevenly across different regimes and places. Approaches such as cultural planning have emerged in part as a response to maintain the distribution and access to a range of cultural facilities and the decentring of mainstream and institutional culture, but even here, this has been marginal through satellites, outposts and franchises of the *real thing*, with cultural and technological power still largely retained at the centre.

More participatory engagement in cultural development and spaces for exchange and experience, on the other hand, has built on earlier community arts movements and more experimental provision such as Arts Labs and community-based arts-led activism. State and municipal cultural provision has also been supplemented and in some respects displaced by commercial entertainment (Adorno and Horkheimer 1943) and more atomised digital consumption, leaving cultural space challenged, but also fuelling a thirst for the authentic and live(d) experience. Responses in the form of immersive and hybrid digital-hyper-real experiences and futuristic but sterile cathedrals of art are particular phenomena (more legacy than *lifeworld*), but these spaces are not where you learn to paint, play a musical instrument or to dance, play or sing collectively, or indulge in traditional and new arts and cultural

DOI: 10.4324/9781003216537-9

practices. The place and process of engagement and experience therefore retain their primacy in both cultural democracy and development, where

> socially engaged artistic or creative practice aims to improve conditions in a particular community or in the world at large . . . instead of a product, socially engaged artists focus on process – one that is designed to affect change. They do this, not alone, but in collaboration with others – and in a public space.[1]

By attempting to put the 'public' back into public art, Lacy coined the term 'new genre' public art (thus distinguishing it from traditionally commissioned public art practice), to represent activist art practices:

> for the past three or so decades visual artists of varying backgrounds and perspectives have been working in a manner that resembles political and social activity, but distinguished by its aesthetic sensibility. Dealing with some of the most profound issues of our time – toxic waste, race relations, homelessness, aging, gang warfare, and cultural identity.
>
> (1995, 19)

In Brazil, two earlier examples of participatory arts practice engaging with social and political issues include the participatory spect-actor-based *Theatre of the Oppressed* and the art-music fusion *Tropicalia* movement. However, artist activists are only one example of socially engaged practice, with other cultural disciplines working in co-designed/produced ways and on issue-based projects with host communities, notably community architects, planners and designers (Evans 2018b). Another sphere of engagement has been in health, drawing on the tradition of therapy – art, drama, music in care home/healthcare settings – and Healthy Living Centres, non-clinic/non-health-based spaces combining arts and well-being activities with public health education, challenging the biomedical model of health (White 1998, 9). Art and other creative skills may well be employed, but aesthetic outputs may not form part of the interaction (Thompson 2012):

> social engaged art that is collaborative, often participatory and involves people as the medium or material of the work. . . . The key element of socially engaged art is the actual participation or experience as opposed to the work itself (if any is produced at all). The act itself – the social engagement – IS the art.[2]

The semantic distinctions arising in this field – community arts, new public art, activist art, dialogic aesthetics, socially engaged art – reflect both the 'social turn' (Bishop 2006) and the adoption of the arts in social policy, but they also mask a disquiet among artists about the professionalisation and depoliticisation of community arts (Braden 1977; Kelly 1984), their 'journey from outsiders to activists to decision-makers' (Merkel 2009),[3] and the commissioning process itself, typically led by funders and intermediaries. The particular position of the artist – independent, freelance, precarious and seeking/protecting their autonomy – has also been considered from the perspective of cultural labour. Belfiore for example criticises the conditions within this practice pointing to the considerable hidden costs (well-being, financial) and invisible subsidy on the part of the artist/creative practitioner and the lack of care offered by commissioning agencies (2022), whilst Banks goes further arguing for 'creative contributory practice', a concept to redress the imbalance between those occupying and those not able to access cultural and creative jobs (2023),[4] an inequity that is often implicit in social engagement between artists, academics and collaborating communities (Brook, O'Brien and Taylor 2020). This field and narrative is however occupied by professional artists and academics in particular, where there is also ambiguity, with artists-as-academics (many artists also work within University Art Schools as lecturers/researchers) and course programmes in community and socially engaged arts practice, and with academics running funded research projects with commissioned artists and local collaborators – the common action–research partnership model. Despite these critiques and positions, there are very few voices heard from those who are actually 'engaged' and the supposed beneficiaries of these creative interventions, or the spaces they occupy and create.

However, as more experience accumulates, knowledge is disseminated and recognition is given to socially engaged (not limited to 'art') practice, particularly collaborations between researchers, artists and host communities, the concept of *co-created*, *co-designed* and *co-produced* cultural space has crystallised. This has found traction in response to urban change and meta-challenges, for example the environment and climate change, regeneration (particularly around cultural heritage and identity), social injustice, and their impact at a local and everyday level. Here, culturally inspired engagement with local communities and communities of interest has articulated creative responses and presented an opportunity, at least, for a power-shift between the conceivers of cultural space and those whose space is lived and represented, raising the question for artists/activists: 'does our work unsettle unequal power relations or does it confirm and support the status quo?' (Moriarty 2014). This challenge is at the heart of much socially engaged cultural practice today, but the notion of *bringing the art to the people* and a presumption of a binary professional-amateur relationship is no longer tenable (Hope 2017). Kester's 'dialogical aesthetics' therefore 'requires that we strive to acknowledge the specific identity of our interlocutors and conceive of them not simply as subjects on whose behalf we might act but as co-participants in the transformation of both self and society' (2004, 79). In practice, engagement around the production of cultural space – physical, symbolic and imaginary – takes place at varying scales, micro and macro – and therefore with different levels of intensity, requiring on the one hand a spatial dimension and awareness, and on the other hand, a closer identification with place and the 'particular' (Wallerstein 1991). The former can be captured through the emerging process of mapping culture, considering people and place in both a spatial and social sense, with engagement centring on this relationship, on local knowledge, and barriers and aspirations towards access to cultural development and spaces.

Cultural Mapping – A Socio-Spatial Approach

The recent concept and practice of cultural mapping (Longley and Duxbury 2016) can be seen as a response to the need, first, for improved distribution of and access to cultural provision and, second, for local knowledge on everyday lived arts and cultural amenities and experience. This encompasses the more strategic cultural planning scale, and the aspirations that emerge from below, through participatory mapping and arts/action research (PAR) and associated visualisation techniques at local-area level. A spatial approach also counters the disconnect between place, cultural provision and participation which is evident through 'placeless' cultural activity surveys (Evans 2016b) and audience data that tells us little about the actual experience (above a price/cost relationship and basic user profiles) – or about 'non-users' (other than their 'absence'). The paternalistic view of users (and critically non-users) of cultural and other community facilities has been long established in the essentially normative provision of civic amenities, and in the design and planning professions themselves. Here, users signify occupiers, inhabitants and in a transactional sense even clients – 'those who would not normally be expected to contribute to formulating the architect's brief' (Forty 2000, 312) – but all suggesting a powerless even disadvantaged role for the faceless 'user' and the problematic, ungrateful 'non-user'. Lefebvre had recognised this tension: 'the word "user" [usager] . . . has something vague – and vaguely suspect – about it. "User of what?" one tends to wonder. . . . The user's space is lived – not represented (or conceived)' (1991, 362). As Forty also observed:

> [T]he decline of interest in the 'user' and 'user needs' corresponded to the decline in public-sector commissions in the 1980s. Perhaps another reason for dissatisfaction with the 'user' has been that it is such an unsatisfactory way of characterizing the relationship people have with works of art and architecture: one would not talk about 'using' a work of sculpture.
>
> (2000, 314)

Whilst national distributive cultural initiatives have declined politically, since the 2000s, the practice of cultural planning (Evans 2001, 2008) has been widely adopted and applied in the strategic development of cultural activities, facilities and resources for incumbent and new communities, particularly in Australia, Canada and the UK (Evans 2008). These have produced more systematic approaches to capturing cultural assets, in particular in response to regeneration, major events, population growth and diversity. The theoretical basis of cultural mapping in particular draws on the 'spatial turn' in the social sciences, including unpacking the idea of socio-cultural space (Duxbury and Redaelli 2020) and place identity (Santos and van der Borg 2023). From the perspective of *Rights to the City* and of equitable cultural opportunities and notions of cultural democracy, a detailed understanding of the distribution, access and 'ownership' of cultural spaces is also fundamental – not to be considered as primarily an administrative or planning concern. Legalistic and political-theoretical claims for *rights* will not alone (if at all) redress the imbalance and deficits in cultural resources which large parts of the population experience. Involving communities in the process is obviously key, although seldom practiced.

Cultural mapping is first a methodology and a set of techniques drawing on various cartographic and digital data analysis and visualisation tools. The development of cultural mapping and planning models has been applied in a number of case-study areas undergoing cultural infrastructure strategies, including areas experiencing population growth and land use change, such as new housing and areas subject to environmental risk (e.g. flooding/erosion), and major redevelopment and regeneration, including tourism development (Santos and van der Borg 2023). The latter scenarios incorporate the role and intervention of practicing artists in visualising and narrating urban change as a consultative and scenario-building process, both complementing and challenging traditional environmental agency/scientist/planner hegemonies – those that conceive the production of space (Duxbury, Garrett-Petts and Longley 2018):

> Artistic approaches are a form of socially engaged practice that highlight critical interest in mapping cultures and the creativity of places, in an attempt to tackle issues of artistic production, and social engagement in the strategic planning of places, stressing the role of artists (and arts) as agents for enhancing community self-knowledge.
> (Santos and van der Borg 2023, 7 and see Edizel-Tasci and Evans 2021)

Cultural mapping, as a stand-alone exercise and resource, or as part of a wider cultural planning and needs assessment process, presents a flexible approach to capturing a particular community's cultural assets, needs and aspirations. This is underpinned by a set of techniques that range from the more systematic cultural audit, consultative planning and visualisation models (Evans 2008), to artist and community-led mapping projects that can engage community creativity, resistance movements and practice-based arts interventions across art forms. However, according to an international review of cultural mapping and guidance (Evans 2008), what constitutes 'cultural assets' in both official classification and mapping exercises varies. Few include natural heritage or environments for instance, whilst some projects are more inclusive in capturing community assets, local heritage and user interpretation of these through local knowledge (Geertz 1985). On the other hand, more sophisticated spatial models have also been developed to plan for changing (i.e. demographic, age, ethnicity) and growing communities and population groups, as well as their future cultural and social amenity needs. This has also seen a convergence of the cultural with sustainable development policy goals, as a form of managed community cultural growth. What this also confirms is that cultural mapping does not draw on a single model (i.e. one size does not fit all) but that it is both socially – and politically – produced (Gray 2006) and reflects national/regional planning and cultural policy systems and priorities (Guppy 1997).

A novel attempt to classify and map cultural assets not just functionally, but in terms of a regional cultural development strategy, was adopted for instance in the city of Toronto, Canada. In formulating the methodology, a wide spectrum of facilities with very different user group expectations was considered. In its broadest sense, since any place where cultural activities might occur, the term becomes unquantifiable. For this mapping exercise, a cultural facility was first identified as a venue that is a building or designed landscape that fulfils a defined cultural role. The term 'cultural facility' was divided into four defined roles, which support specific municipal objectives and responsibilities related to culture: (1) support for cultural activity throughout all of the City's diverse communities; (2) support for the artists of the City; (3) support for culture as part of the City's Economic Development and Tourism strategy; and (4) support for culture as a heritage resource. These roles were then described in terms of four categories in which individual cultural facilities and organisations were represented: *Hubs* provide support for cultural activity throughout all of the City's diverse communities. They tend to be community driven and nurture cultural activities at a local level. About 60% of hubs tend to be concentrated in the downtown core and about one-third are City owned. *Incubators* provide support for Toronto's artists. They tend to be artist-run facilities, heavily clustered in specific urban neighbourhoods; and *Showcases* provide support for culture as part of the City's Economic Development Strategy – these facilities often have regional, national or international profile (Evans 2008). Over 750 cultural assets were identified and mapped in this cultural asset exercise (Davies 2003).

North Northants Living Places

As an example of regional-level cultural mapping, a Cultural Infrastructure Plan was developed for North Northamptonshire (Northants) in central England – a designated growth area requiring investment in new and upgraded cultural facilities and improved access in a subregional area with no major metropolitan cities and therefore no higher-level facilities. This exercise was also used as a pilot in the creation of a national Cultural Planning Toolkit (Evans 2008; DCMS 2010). Comprehensive mapping was undertaken, with over 25 detailed digital data maps across cultural, environmental and social domains, in collaboration with local authorities, a development agency, a regional arts organisation and other cultural bodies. The context was that of a growing population and specific housing growth areas, as well as town centre regeneration (e.g. Corby) in what is a mixed post-industrial (e.g. steel manufacturing[5]) and semi-rural region, with a socio-spatially divided population. Extensive baseline mapping of a range of socio-economic distributions included household income, educational qualifications, population density, age ranges, disability/illness, and lifestyle groups – all potential indicators of cultural participation and 'cultural capital' – along with population and housing growth over the following 20 years. The categories of cultural amenities are indicated in the example map (Figure 9.1), in which they were 'layered' over the various spatial data analysis and housing growth areas where cultural facilities were likely to be most needed. These annotated maps were used as the basis for consultation with residents and stakeholders and to highlight the distribution of cultural assets and gaps in access and provision. For example, regeneration-led cultural

Community Scale Culture Facilities

North Northamptonshire Cultural Infrastructure

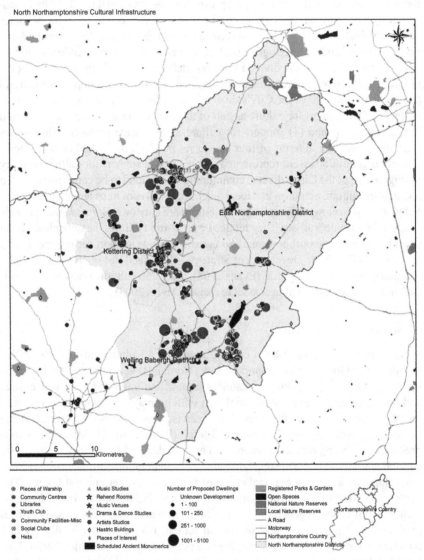

Figure 9.1 North Northants Community-Scale Cultural Facilities and Growth Areas

facility development included a newly built *Corby Cube*, combining library, health centre and other town centre facilities with 400-seat theatre, studio/lab and basement media/recording studio, but the town lacked a single cinema screen, as was evident from the mapping and consultation. Furthermore, the 'rational' relocation of a youth theatre to an exhibition centre, away from the concentration of young people, local transport and the town centre of Kettering, also emerged from corre-lating population groups with amenities and accessibility. Both cases revealing the tendency to build the wrong cultural space, or in the wrong place (Evans 2015a).

Engagement also included a residency-by-community arts organisation *Think Space* within the town, working with local residents on a range of local issues/themes and routes, through artworks, events and other interventions. These included *Construction/Destruction* with photographic images displayed on hoard-ings by building/demolition sites throughout the town; *Vision Pub Space*, an instal-lation in a disused pub/nightclub with images used as a starting point for discussion with residents exploring issues about experiencing change and re-use/regeneration of the town centre; *Rest and Play* a series of 50 signs located throughout the town followed by a market stall to talk to people where they rest and play (mostly not in the town centre environs); and *Changing Spaces* a focal point for communi-cating with young people how visual art created by the community could make the townscape more vibrant and stimulating, with workshops and installation aris-ing; as well as *Debate days* for local artists and residents. This groundwork and the detailed layering of cultural map data informed the strategic cultural plan for this region (Fleming 2010) at a local-area level and provided an historical record and reference against which the realisation of cultural space development could be compared and measured over time.

Shaping Woolwich Through Culture

At a more local rather than regional scale, *Shaping Woolwich Through Culture* worked with detailed facility-location information to enable accurate cultural asset identification at a fine grained geographical scale. This level of detail increased the analytical potential of the information and its use in an area-based approach to developing a strategy for the town centre. In Woolwich, located in south-east London, where a key driver was the support for cultural infrastructure development in areas of anticipated housing growth and population change. Further analysis of the accessibility of existing cultural assets also helped to identify gaps (quality and capacity) in both current and future provision, after the new housing development was completed (see Figure 9.2), as in the preceding case of North Northants. In Woolwich, knowing the relationship between individual development sites, pro-jected population growth and existing cultural asset locations was considered criti-cal to building scenarios for the creation of Woolwich as a 'good place to live and work'. Analysis of the spatial clustering of physical assets also led to the identifica-tion of 'cultural nodes'. It is also possible to annotate visualisations with data from an inventory to display information about the size, quality and use of individual assets (e.g. capacity/number of seats, technical specifications). Cultural mapping can also employ visual consultative methods such as GIS (Geographic Information Systems)-Participation (GIS-P) with small groups working with large-scale maps that can be annotated with perceptual, as well as experiential feedback, includ-ing walking times to local venues (Figure 9.3). This local knowledge and senti-ment can also be digitised back into interactive maps containing geodemographic,

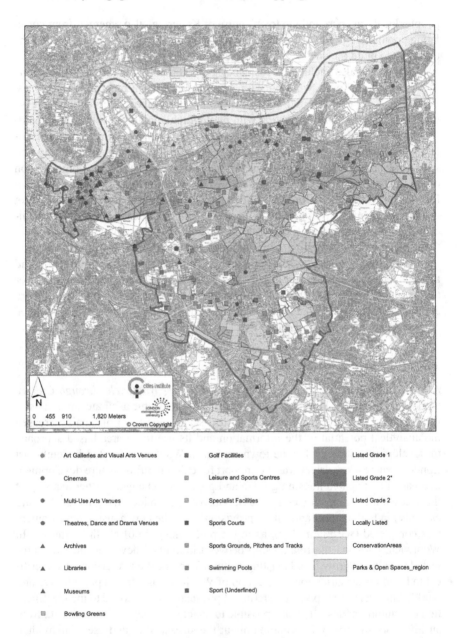

●	Art Galleries and Visual Arts Venues	▣	Golf Facilities		Listed Grade 1
●	Cinemas	▣	Leisure and Sports Centres		Listed Grade 2*
●	Multi-Use Arts Venues	▣	Specialist Facilities		Listed Grade 2
●	Theatres, Dance and Drama Venues	▣	Sports Courts		Locally Listed
▲	Archives	▣	Sports Grounds, Pitches and Tracks		ConservationAreas
▲	Libraries	▣	Swimming Pools		Parks & Open Spaces_region
▲	Museums	▣	Sport (Underfined)		
▣	Bowling Greens				

Figure 9.2 Woolwich Cultural Facility and Heritage Assets

Figure 9.3 Multipurpose Visual and Performing Arts Venue Catchments

facility, transport and other data and be repeated iteratively with the same or different groups (DCMS 2010). This technique, which draws on the earlier *Planning for Real* model using simple board games, models and maps, is utilised successfully by users from primary school children to pensioners, and around urban design, transport and heritage interpretation (Evans and Cinderby 2013; see Chapter 4), as well as in conflict sites and resolution situations.

Natural Cultural Districts

In another area-based approach, the concept of Cultural Districts has been used in the USA, defined as a well-recognised labelled, mixed-use area of a city in which a high concentration of cultural facilities serves as the anchor or attraction (Stern and Seifert 2007, 2010), and which are often centred near to large arts institutions. The idea of *Natural Cultural Districts* was developed in Philadelphia, in areas suffering multiple deprivation, population and economic decline, building on the idea that urban neighbourhoods often germinate clusters of community, commercial and informal cultural assets linked by artists and creatives as producers, and participants as consumers or practitioners. The study used four indicators of the intensity of the cultural scene in a neighbourhood: cultural participation, non-profit cultural providers/community associations, commercial cultural firms and independent artists/creative workers, together making up an area's cultural assets. Four sources were used: a regional inventory of non-profit cultural resources, a database of commercial cultural firms in the metropolitan area, a listing of artists, and small-area estimates of regional cultural participation based on data provided by over 75 cultural organisations. All four of these indicators were calculated for every census block group (approximately six to eight city blocks) in metropolitan Philadelphia. The identification of natural cultural districts used factor analysis to create a single scale, capturing variation of all four of these indicators across the metropolitan area. The analysis determined that the four indicators had very similar patterns of variation (a single scale accounted for 81% of the variation producing a *cultural assets index*). The second stage identified neighbourhoods with an index score higher than expected when corrected for these variables such as socio-economic profile, diversity and distance from centre. Essentially, these are districts that were 'exceeding expectations' in their concentration of cultural assets. In order to test the role of cultural spaces in neighbourhood revitalisation, the model combined the cultural assets index with data on neighbourhood change. The results were striking: 83% of all block groups that improved by two or more market value analysis (MVA) categories between 2001 and 2003 were cultural districts and this association between these clusters and improvement in housing market conditions has continued (Evans 2016b).

Site-Based Practice

Whilst area and spatially produced cultural plans provide some basis for engagement, inevitably they suffer from a birds-eye and top-down perspective. Despite advances in community-based mapping using GIS-Participation and the transfer of visualisation skills and technology to local people, they can often be over-reliant on technocratic intermediaries (academic, local authority, cultural agencies and consultants) unless combined with a more dialogic relationship which requires the creative co-design and co-production of the cultural spaces and experiences that might be produced through social engagement. Site-based artist involvement is perhaps the most familiar scenario in socially engaged practice, often coalescing around specific themes, changes (physical, economic) or even threats to the status quo, and this is generally distinct from the outreach work emanating from established cultural centres and their involvement with local communities. Examples of site-based engagement include major regeneration projects in PobleNou *@22Barcelona* and Wangsimni New Town in Seoul (Kriznik 2004), with artist and community-led intervention in the development process literally in their own backyards. Despite these initiatives, in both cases declining social cohesion and a lack of citizen participation were found to be a consequence of speculative urban development, in which urban regeneration was instrumentalised to attract investment and improve the city's global appeal rather than address endemic local challenges. Visible resistance is also evident in areas undergoing gentrification, notably through street and graffiti artists working in collaboration with local communities as discussed previously. The following examples of socially engaged cultural space production in everyday areas of the city provide some insight to both the process and scope of involvement, from the seemingly small-scale artist activism and creative community media, to complex environmental and heritage scenarios involving multi-disciplinary creative teams working in participatory arts and action research. All represent cases of embedded practice and incremental collaboration with host communities outside of institutional settings and a move towards co-design and co-production, in contrast to the short-term intervention (e.g. artist residencies, public art commission) and culture used in the service of redevelopment where 'each story of regeneration begins with poetry, but inevitably ends with real estate' (Kunzmann 2004b, 2).

I Fail to Agree

In Sheffield, West Yorkshire, for instance, artists were directly engaged in the process of the redevelopment of the city – in particular the gentrification of Devonshire Quarter, a new housing development project. Andy Hewitt and Gail Jordan are site-based installation artists who had a studio overlooking Devonshire Green near Sheffield city centre. Two projects were commissioned and undertaken by this team, both focused on the Devonshire Quarter area: *Outside Artspace* and *I Fail to Agree: Hewitt & Jordan* (Beech 2004). In *Outside Artspace*, the artists worked with the city planning department to help develop a vision 'to reinforce the identity of the area and improve land use, transport, urban design, the local economy, housing mix, sustainable living, quality of the environment and community safety' (ibid., 26). This area has been associated with youth activity and small businesses serving this market (skateboarding, record shops and cafés) and a growing university student body, due to the development of new halls of residences nearby (Evans and Foord 2006b). During this process, West One, a large-scale, eight-storey apartment development, was under construction overlooking the only large green space left in the city centre. The artists visited the West One showroom to discuss their vision for the development. They said that the council planned to build a bandstand, create a pleasant safe area with CCTV – an image directed at the 'exclusive' apartment market, with the green as a 'front garden' feature for new residents, rather than a as community, social and public space. The artists' proposals arising from community consultations included a venue for art projects, exhibitions, film, performance, music events – as part of an annual programme – and youth facilities, including a skateboard park. These proposals were received by the Council and contact with them then stopped – the recommendations were not taken up. Five years later, in the planning consultation exercise, led by commercial masterplanners EDAW, concern was expressed that the Green skate park was not shown on the (new) plans and they had heard that it was being got rid of. Forms of dialogue and engagement proved to be merely cosmetic, in this case, a familiar exercise in co-optation.

Whose News Is It Anyway?

Resistance is not confined to local artists of course, but local communities have expressed their anger at the so-called culture-led regeneration process and so-called housing renewal, through community newspapers. For example, as reported by the community paper, *Salford News*, the city of Salford – 'poor cousin', literally, to its well-heeled Manchester neighbour – hosts Salford Quays, a 1980s' redevelopment and central government-inspired (and funded) urban regeneration zone, now hosting the Lowry Arts Centre and Imperial War Museum of the North, and relocated BBC studios at a new *Mediacity UK* development, adjoining Salford University campus (Christophers 2008). This new cultural quarter is served by an extension to the Manchester Metro Light Rail – but this transport link does not go to Salford town centre itself, or to where most local people actually live, including young people who have little or no 'ownership' of the arts complex, from which, not surprisingly, they feel excluded.

In 2006, six local lads eponymous 'hoodies' from an east Salford estate were asked by the *Salford Star* newspaper to visit the centre to see the Lowry painting exhibition (depicting local factory workers and 'working-class people off similar estates'). On a wet Sunday afternoon they entered the building, went up the escalator to the exhibition and walked past the information desk, into the gallery. They were stopped within two minutes of entering this 'free' venue and refused entry. Security was called, but no reasons were given for this by the staff. Another visitor at the time commented: 'basically they were local lads coming in to look at the pictures because they were bored stiff and they were denied access to a facility which we've been told is open to everyone' (Evans and Foord 2006b). Such stories would go unreported without community news, which would otherwise leave such interactions as anecdotes to the myths perpetuated about the divide in cultural capital that culture-based regeneration inevitably accentuates.

Docklands Community Poster Project

The use of print media, newspapers/letters and posters was a device used extensively in the low-tech era of the 1970s and 1980s in response to large-scale city redevelopment. In London's Docklands, wholesale regeneration of this area was met with strong local resistance. The local trades' council approached local artists Peter Dunn and Loraine Leeson to assist residents in their campaign – resulting in the creation of the Docklands Community Poster Project (DCPP) in 1981:

> They asked us to work with tenants and action groups to produce a poster alerting local people as to what was to come. Tenants in the area were already federated into local action groups and when we consulted them, thought posters were indeed wanted, but large ones to match the scale of the proposals.[6]

Dunn and Leeson's summary captures this now historic scenario: with miles of corrugated iron surrounding what was left of the docks, local people were left stranded in poor housing with few facilities, and unaware about what was going on. So the main audience for the proposed large-scale posters, or 'photo-murals', would be the Docklands communities themselves. The images were mounted on billboards 18 feet long by 12 feet high. The siting of these images was also important since, unlike commercial advertising billboards, they were positioned where they could be viewed by pedestrians rather than passing cars. The first was installed opposite a health centre, then over subsequent years there were seven more around the Docklands area, six operating at any one time. The photo-murals each comprised 18 sections, mounted onto plywood panels. In this way, the murals were able to change gradually through replacement of individual sections, and develop a narrative, like a slow motion animation. In practical terms this meant the images could be transferred from one site to another, enabling the story of the Docklands to unfold through time and space – *The Changing Picture of Docklands*.[7] The Docklands Poster Project continued for ten years until 1991 (Leeson 2018) reflecting the degree of both engagement and commitment of the artist team and organisation with their collaborating community and place. Leeson in particular has continued this place-based engagement with successive longitudinal projects (see *Active Energy* cspace.org.uk).

Hornsey Town Hall – Community Futures

As discussed in Chapter 4, attitudes towards cultural heritage – both official, designated and non-designated, vernacular – vary. Much of what might be seen as commonplace heritage including that situated within everyday environments, maybe be considered benign, or not even special. Until, that is, they are threatened in some way (culturally or physically, including change-of-use), when historic buildings can act as focal points around which communities will rally and revive their sense of civic pride (OBU 2003). As Castells puts it: 'the symbolic marking of places, the preservation of symbols of recognition, the expression of collective memory in actual practices of communication' (1991, 351) are very important in order to recognise and, if necessary, protect the identities of places. However, the cultural assets that communities value are not always the same as those that local authorities consider as 'culturally' or environmentally 'significant' (Cauchi-Santoro 2016).

The model and precedent of the arts centre present a particular opportunity for both responding to current and prospective community and cultural needs, and for a sensitive adaptation of a listed building which can be flexible and viable over time. Any mono-usage – whether cinema, theatre, hotel or housing – effectively closes off these possibilities and is less likely to reflect the needs of resident or cultural communities. The arts centre conversion is also a proven re-use option. As discussed in Chapter 2, they are architectural opportunists with over 80% of arts centres in the UK housed in second-hand buildings, and as well as flexible and adaptable use of space, the temporal dimension offers the opportunity for complementary usage at different times, and also evolving over time – during temporary 'meanwhile' use, and as the former Town Hall in this case re-establishes itself as a community resource which will change as cultural and other uses emerge. The preference for choosing existing civic structures for cultural use is no accident, nor just a case of economic and political opportunism. Their symbolic importance is also based on a deep history of the sites concerned with previous usage, buildings and often sacred and even profane significance. This can be captured in the notion of a sense of place, not just embodied in the building, aesthetic and spatial features. As Norberg Schulz suggests:

> [T]he spaces where life occurs are places. . . . A place is a space which has a
> distinct character. Since ancient times the genius loci, or spirit of place, has

been recognized as the concrete reality man has to face and come to terms with in his daily life.

(1979, 5)

In the 1920s, a local north London council had bought the long, wedge-shaped site of the present Hornsey Town Hall. It contained Broadway Hall (destroyed by fire in 1923), Lake Villa and some cottages. The Council laid it out as a public park with a playground. By 1929, the Council had a plan to build their offices above the Broadway frontage, subsidised by shops below. However, its main civic functional use lasted a little more than 30 years, from when its decline as a municipal and occasional cultural centre gradually commenced, which accelerated as the Council's occupation reduced whilst the building's fabric and structure deteriorated. Interest in saving and adapting the Town Hall has been a growing concern during this period of decline and neglect, with several attempts at design schemes raising expectations, only to be dashed each time. The council-led proposal was to effectively sell the building and public square to a foreign property company on a 125 year lease, with a private housing development at the rear of the building/car park and with housing also encroaching into the historic building itself (Grade II-listed). Surprisingly it has been the current period of temporary occupation that has seen the latent demand and focus for a range of community, cultural and business uses – over 100 cultural and community organisations, including a pop-up HTH Arts Centre organisation, an art gallery focusing on local and under-represented artists, as a location for film shoots, weddings (as the former town hall also provided) and the annual Crouch End Festival. New community organisations and networks have been formed arising from this prospect and perceived threat from the privatisation of this community asset.

    In order to coordinate the community and cultural sector's response to the re-use options and Council proposal to effectively sell-off the building and square, a Preservation Trust had been formed by members of the Crouch End Festival, resident groups, including the resident Ply Art Gallery, representatives of the local university's Art and Design school and the local Historical Society with two things in common, they were all local residents and shared a vision for the community cultural use of the space. Artists and designers (architects, interiors, engineers) were thus part of this community and also had access to resources – funding, students, researchers, design studios, etc. Funding in this case became available via the Arts & Humanities Research Council (AHRC) through a small festival grant as part of a national *Community Futures* programme celebrating 500 years since the publication of Thomas Moore's *Utopia*.[8] The promotion of an open exhibition at the Hall also coincided with the annual London Festival of Architecture which ran annually each June. Held over ten days *Community Futures and Utopia: Yesterday, Today, Tomorrow – use and re-use strategies for Hornsey Town Hall* received over 3,500 visitors. In the co-design stage, a series of community consultation events were held by the Trust, open to the public, at which schemes for the re-use and council proposals were discussed. Demonstrations were also organised in the public square (Figure 9.4), including market stalls and street performances and exhibition at the

Figure 9.4 Hornsey Town Hall, 'Not For Sale'
Source: Photograph by the author.

local library nearby. The intensive engagement centred around a design exhibition undertaken with local interior architecture students and exchange staff and students from Politecnico di Milano (Polimi), all of whom had access to the site, including redundant areas. The design proposals were developed iteratively through resident consultation events – meetings and on-street, oral history workshops, co-design charrettes and feedback with residents, with a number of design schemes exhibited including drawings/elevations, models and videos at the Ply Gallery over ten days, with visitors encouraged to give their feedback and their own ideas on the re-use and current proposals. As a former civic centre, local resident's memories were evocative with everyday usage (registering births and deaths, paying taxes, parking fines, viewing planning applications), mixed with the formal (attending council meetings, weddings) and the memorable, notably concerts including 'up and coming' local bands – The Kinks, Queen and the Pretty Things.

In the Hornsey Town Hall as part of the exhibition, two feedback walls were also created in which participants were encouraged to respond on *post-it* notes to a set of questions on a printed postcard handed out at the entrance and to give their general feedback on the exhibition design themes and overall experience of the place (Figure 9.5). The consensus confirmed that the public wished to see this heritage building retained for community and cultural use with many innovative ideas for specific usage and activities. At a follow-up meeting with the local Council, staff were very positive about the project and its reception. The Councillor for Culture was also very supportive and expressed a wish to develop a more strategic

Figure 9.5 Hornsey Town Hall Exhibition Feedback Walls
Source: Photograph by the author.

and linked way of working with this community in the future. The refurbishment of this cultural asset was set to degenerate into a private housing, hotel and club venue; however, the engagement with local artists, residents and cultural organisations was able to articulate the strength of feeling and visions for an accessible cultural space. The mixed-use development, effectively saving this deteriorating heritage building and square, has ensured the Assembly Hall, co-working spaces and smaller venues for performance and workshops are retained with access hopefully ensured for community groups, schools and local artists.

DEN-City

The use of festivals as an engagement opportunity ranges from community festivals co-produced with professional and local participants, leading to community plays and music performances which may be designed around local themes (Moriarty 2004), to community-based events which explicitly explore social and place-based issues through the creation and curation of cultural spaces (Chapter 3). Engagement of this type which involves local artists and cultural organisations, also reflects their perspective on these shared issues and where the distinction between cultural and community actor is blurred. Presence in a locality or neighbourhood may indicate shared space and common interests, but distinctions may also persist between these groups which may be class-based or where their experience of the everyday may still differ. The assumption that a community in a given area or neighbourhood is uniform and does not contain variations in socio-economic, housing, employment, prior knowledge (e.g. incomers/new and established residents) and cultural and social capital is obviously false.

Participatory action research (Pain et al. 2012) therefore requires issues of representation, capacity and a process of trust-building to be pre-requisites to collaborative working, and the co-design/co-creation of any cultural initiatives arising. Prior to the event in this case, cultural mapping sessions were held with local residents, including elected local councillors, and attendance at monthly meet-ups with local organisations, artists and tenants association representatives over the period of a year. Mapping at open festival environments also brings together locals and visitors who may have differing perspectives and experiences of the area and its valued cultural assets, whilst also providing the opportunity for participatory arts activities, installations and performances to allow cultural expression, exchange of ideas and complement the responses to the cultural map, and vice versa. Unlike the more closed group meeting (e.g. community consultation events, focus groups), arts and community festivals can also bring local residents and users of a neighbourhood together in a more friendly, relaxed and animated way. Here, the conversation through cultural mapping helped to identify not only tangible aspects of the area but also intangibles such as stories, histories, values, 'the aspects that provide a "sense of place" and identity to specific locales, and the ways in which those meanings and values may be grounded in embodied experiences' (Longley and Duxbury 2016, 2). The dialogue developing around the physical map captures features that are not always easy to quantify, but are important to truly understand a place and its value to its residents and visitors. In this case, issues raised through initial mapping and engagement included pollution (water, noise, air, waste, construction), the lack of accessible cultural amenities and open space, the loss of work (and live-work) space for artists, and the shortage of affordable housing and 'crowding out' through the development of new up-market housing. People's interactions in their community as well as personal and collective memory helped to build the narratives. The activity itself helps to create a community-driven 'visual' of values and place-based meanings which are evidently different from official plans and maps (Cauchi-Santoro 2016) and even of official history, narratives and worldviews. This is in contrast to the master-planning process which dominates the design and development of major regeneration sites, as experienced by these communities as a result of the major redevelopment of the wider area during and following the nearby London 2012 Olympics site (Evans 2015b).

The community festival, held over a long weekend in June, encompassed diverse events, exhibitions, workshops and talks at community and cultural venues in the local neighbourhood, including an exhibition of design projects based on waterside heritage buildings, held at the local youth centre. A cultural mapping stand was mounted at outside cafes and workspaces with over 40 visitors participating in the mapping exercise, whilst during the evenings there were screenings of the East End-based *Long Good Friday* film at the Mother Gallery, as a backdrop to how this area has been popularised – *viz.,* ganglands, industrial decline and docklands redevelopment and extreme gentrification – and open debates were held on topics such as community land trusts and artists workspace. During the festival, alongside artists' open studios, street performances and design exhibitions, a derelict site was occupied along the canal to build a temporary *DEN-City* from recycled materials and rubble where residents and visitors could explore the waterside environment in the context of urban change and sustainability with a group of independent artists, whose installations, artworks and performances reflected and responded these concerns (with artists using waste/recycled materials in their outfits – Figure 9.6). DEN-City was curated by local artist Rebecca Feiner at Forman's Yard Fish Smokery, a vacant plot in the shadow of the Olympic stadium. *DEN-City* according to Feiner is a temporary utopian city of installations, dens and assemblages. Colourfully, repurposing, and recycling on the theme *Work in Progress*. The event was captured on film by environmental film-maker Sara Penrhyn-Jones,[9] and discussions arising from the event were broadcast on community radio.

In many respects, these community-focused arts festivals follow a pattern of slow initial growth, then peaking after which energy and reliance on a small group of organisers and artists fades, unless a sustainable strategy (often funding) and exit strategy is place. This also reflects the changing community and context within which the festival is situated – formed in response to a need or gap in provision, for local voices to be heard, and local issues aired. Where these resolve or change, the original purpose also changes and the festival rationale may also change or disappear altogether (but also leaving open the possibility for resurrection at a later date). What is important here is the long-term presence of site-based artists and arts groups who are *of* the community and whose practice is embedded in place.

Three Mills Heritage and Active Energy

Three Mills represents a complex of heritage buildings including the world's largest water mill wheel built over the tidal section of the River Lea, a tributary of the

Figure 9.6 DEN-City Festival, Fish Island
Source: Photographs by the author.

Thames near Bow in East London. Neighbouring buildings include an academy science school and at Three Mills Island, a complex of studios used for film, music rehearsals and recording. The Trust responsible for the House Mill, a Grade II* listed building, was seeking possible re-use options for the House and to increase awareness of the site and its history and heritage value, particularly among local residents and schools. The site regularly flooded due to upstream damming and a high tidal flow, with pollution effecting fish and wildlife and water quality.

The community engagement activities at Three Mills were undertaken as a part of a three-year AHRC *Connected Communities* research programme with a particular emphasis on co-design and co-production and the concept of cultural ecosystems (Fish and Church 2013). A range of research and engagement methods were adopted as part of an overall Participatory Action Research (PAR) approach, with a focus on citizen science, sustainable design and visualisation, in order to engage in what is a complex environmental sphere and set of challenges. The research project was led by a combination of academics and artists, with architecture and engineering students (BA to PhD level) and a range of local partners including an elders group from the local Age UK branch, the House Mill Trust and residents association.

Three integrated, collaborative engagement activities included: Cultural Ecosystems Mapping; Active Energy; and Re-Use Co-Design. These activities engaged with communities in different ways and at differing scales. Cultural ecosystems mapping ensured a participatory process at neighbourhood scale, including waterways and heritage facilities, and identified the needs and opinions of the community. The Active Energy demonstration water turbine project brought older and younger people together to navigate environmental change in their community, and participants learned about the need for sustainable forms of energy and clean water generation to help counter climate change locally – upstream and downstream – and the wider implications for global warming. Finally, university architecture students selected the Three Mills site for their final-year major design project. These schemes were exhibited at the House Mill which aimed to raise awareness of the environmental challenges and presented possible solutions to developing the waterfront both imaginatively and sustainably for the future. Although the process and depth of creative engagement were at the core, an outcome of the three activities was also to inform local agencies and other policymakers about the values, concerns and knowledge that people have of their environment and heritage, and increase community engagement in climate change policy and initiatives.

Cultural Ecosystems Mapping

Cultural ecosystems mapping first draws on a conceptual framework (Figure 9.7) which seeks to identify particular benefits of cultural assets and experiences, in this case, cultural and natural heritage-related 'services'. Elsewhere, it has proven to be a 'valuable tool to articulate community perspectives, experience and aspirations and thereby to inform local agencies and other policymakers about the values, concerns and knowledge that people have of their environment' (Edizel-Tasci and

Evans 2017, 135). The mapping exercise was able to integrate a number of environmental and quality of life issues at an understandable spatial scale, capturing useful information from participants of direct benefit to the Three Mills heritage venue itself, and raising awareness of future engagement opportunities and aspirations for the site. This included the renovation of the water mill wheel and House Mill which visitors were able to explore over two weekends. Mapping was undertaken outside of the Mill House over several days catching regular users, visitors and the adjoining school. The presence of the local team with hosts over several months and repeated over two years, meant that users of the space became familiar and intrigued with the project as it progressed, particularly school children who passed by several times a day between lessons as the physical design aspect of the project took shape.

Perspectives on local cultural ecosystems were collected initially from workshops with residents held in a canal-side locations, and were later analysed to derive local community values (Ryan 2011). In order to place this site in its wider riparian context, visual artist Simon Read walked the entire length and breadth of the River Lee over several days, documenting and photographing the landscape, its

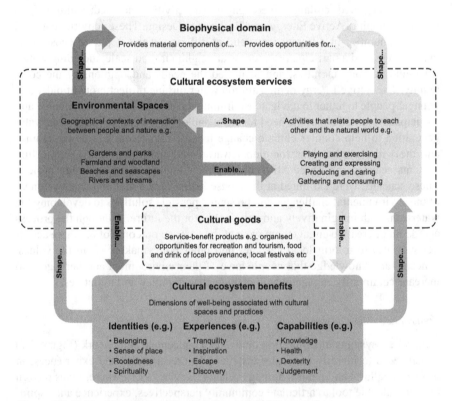

Figure 9.7 Cultural Ecosystems Services Conceptual Framework

Source: Fish and Church (2013, 211).

uses/observing users and physical state (Read 2017). Naming the system *Cinderella River*, themes that recurred from this intensive exercise included access, amenity, conservation, environmental quality and governance. The following aspects were also informed by the Millennium Ecosystem Assessment (Plieninger et al. 2013) and were adapted for the ecosystems mapping exercise with users/visitors. The Cultural Ecosystem Services Framework (Figure 9.7) makes a distinction between cultural values, environmental spaces, cultural practices and cultural benefits. For the initial mapping these were operationalised in terms of *Use*: sense of place, activities, recreation; *Aesthetics*: aesthetic values, spiritual values, inspiration; *Cultural Use*: recreation, social relations, cultural heritage values, knowledge/ educational systems; *Problems*: accessibility, safety, unpleasant; and *Community Cohesion*: diversity, engagement/involvement. The visual results and interpretation arising from the mapping can be found at Edizel-Tasci and Evans (2017).[10]

The cultural ecosystems mapping was also able to validate and support a socially engaged art project, *Active Energy*, which benefitted from co-design and co-production of knowledge and the articulation of community aspirations for the area (below). Findings and suggestions arising from the annotated maps were, in particular, able to inform the design of the *Active Energy* project, in terms of the precise location and theme that the water turbine would address and highlight, as well as provide an indication of the spatial relationships up and downstream of the site. This was important in view of the flows (flood water, pollution, people) and different impacts felt across this waterway. One of the major findings from the mapping revealed that local meeting places where people get together to have a drink/meal or just to enjoy the natural environment are also considered as a part of local and heritage. This showed that people like to spend their time around locations which they value as part of their cultural heritage, but which are under threat in the case of pollution or flood risk due to climate change and over-development. These findings were also shared with local policymakers and NGOs responsible for the waterways helping to raise awareness to climate change at local and city-regional level.

Active Energy and the Geezers

Active Energy first started as a participatory action research project on the democratisation of technology development by a group of local community artists. The team explored 'how the experience of older people was not only being excluded from the development of new technologies, but often left this age group victim to the technological design and control of others' (Leeson 2018, 64). A group of retired men, all former dock and maritime workers from the local area, including an engineer who had worked with steam turbines, and others with mechanical interest and experience, had discussed how water wheels might be used in nearby tidal waters to produce renewable energy. Meeting at a local pensioners club at Age UK in Bow, the idea to generate electricity from a water turbine for a public art installation there, led to the introduction of a local community artist to develop this idea further. This was the start of a long-term collaboration of local artist, Loraine Leeson, with the self-named pensioners group, *The Geezers*. Conscious of the impact of rising electricity costs to those on lower incomes, the group had asked – how could technology be used to harness the power of tidal water as a sustainable source of energy, and which could improve life for themselves and their community? *The Geezers* continued developing their ideas for tidal turbines, making their case for the local resourcing of renewable energy and also to start an inter-generational project at a local boys' secondary school. This was motivated by their desire to pass on their local knowledge and experience to a younger generation and promote the potential of renewable low-cost energy from this local natural resource. This led to *The Geezers*, who were previously isolated older men, finding themselves mentoring underachieving local schoolboys.

The outflow from House Mill at the Three Mills site was designed to be utilised to drive a low-cost/tech-floating stream wheel, powering an aerator to help oxygenate the water and counteract the effects of pollution on the river's fish and wildlife. This riparian heritage site was a microcosm of climate change – downstream of areas undergoing major development and flood prevention measures, with rising water levels and flooding, and water pollution from road run-off and upstream spillage into the river and feeder-canal. The Active Energy mobile water wheel was launched at Three Mills (Figure 9.8), below the House Mill building during the National Mills Weekend, a national event to celebrate mill heritage. This was an opportunity to bring awareness of renewable energy, the importance of community input to design, and the value of the older population to wider society. The turbine remained in situ for a month, during which time birds nested on it, the tide made it rise and fall 20 feet daily, and at one point, it broke loose from its mooring and had to be reattached. Underwater filming revealed fish, bird and other water-life drawn to the motion and aerated water. Its presence made its sighting both familiar and novel, eliciting constant questions from passers-by and visitors to the site.

Following the perceived success of the water wheel project, three years later, a second water wheel based on the original model was created by the same team for the Waterworks River in the Queen Elizabeth Olympic Park upstream from Three Mills. Science students at another secondary school located nearby the river and post-16-year-old engineering students at a college of further education were

Figure 9.8 Mill House, Water Turbine Installation and Schoolchildren, Three Mills
Source: Photograph by the author.

involved in workshops leading up to its installation. Engagement was facilitated through the school and college STEM (Science Technology, Engineering and Maths) curricula, which 'justified' the involvement of students during school time, and also contributed to their wider learning around sustainability and the local environment (note that this was/could not be presented or rationalised as an arts or humanities project). As Leeson (2020) observes:

[P]articipants visited the wheel in situ and learned about the need for sustainable forms of energy to help counter climate change as well as the ecological challenges for their local rivers. In group workshops working models of turbines suitable for the generation of renewable energy were created.

The project team presented this work at the London Legacy Development Corporation (a mayoral agency charged with developing the post-Olympics zone), held during a week of world action for climate justice. The *Active Energy* project celebrated how older and younger people have come together to work for environmental change in their community.

Re-Use Co-Design

Engagement in heritage re-use in the context of climate change and wider environmental issues was also undertaken with local university architecture design students who had selected the theme, after local consultation, of: *Edge Condition* (Evans and House 2017), signifying the water/land boundary, edge city and liminal state of this post-industrial and ancient environment. Students had spent several months in fieldwork, investigating waterside buildings – their past and recent use – as well as related infrastructure and communities of interest, with the challenge to produce design visions and concepts for these sites along this waterway. They had access through lectures undertaken by research team members to the cultural ecosystems findings and visualisations, and as part of the design process had undertaken fieldwork and site visits, talking to local people and organisations about the history and issues surrounding the waterside buildings and their potential use. Following the Active Energy water turbine installation, an exhibition of their final design schemes for the re-use of Three Mills heritage buildings was organised in collaboration with the Three Mills heritage organisation, in order to maintain the momentum and continue engagement.

The opportunity to display their work was also enhanced through the London Festival of Architecture (LFA) which was held annually across the city in June, with the year's LFA theme, *Memory*, which was particularly apposite to the industrial and intangible heritage of the area. A month-long exhibition of student design projects was housed at Three Mills in the foyer/café area. This provided local residents, visitors to the Three Mills Heritage Centre, and a first-time architectural audience, to view and experience the site and thus raise awareness of the environmental challenges, history and possible solutions to developing the waterfront both imaginatively and sustainably for the future. Design concepts displayed at the

Three Mills exhibition ranged from a 'Post-Apocalyptic Flood Survival Centre' *Sinking Future* by Darta Liepkalne, to a *Boat Crafting Station* and various creative and cultural spaces for community and educational use (Evans and House 2017, see Figure 9.9).

These case studies of community engagement in social, economic and environmental change through arts-inspired collaboration highlights the importance of co-design and co-production in research-based interventions that stress the importance of local knowledge (Geertz 1985) in a field that is dominated by complex technical and spatial dimensions, opaque governance systems, as well as time horizons that are beyond most community comprehension, or at least, the power to effect real change. Information provision, consultation and participation have been developed through the engagement activities at Three Mills as in the previous case studies, as well as raising awareness of environmental and key development issues in the area. The value of inter-generational collaboration and the use of socially engaged creative practice, purposeful citizen science (with ownership by participating citizens) and visualisation of community experience and aspirations towards their environment are all valuable lessons. This includes the benefits of accessible technology that is potentially both inclusive and sustainable (Evans 2013a). The use of active heritage and everyday cultural spaces (youth centre, local venues, gallery/shops/studios) as sites for engagement and innovation has also provided a useful context to draw on memory, skills, adaptation and resilience in the face of multiple environmental challenges in areas undergoing pressures from urban development. In practice, all sites have a cultural heritage dimension which may be implicit or need

Figure 9.9 Sinking Future, Three Mills Exhibition
Source: Photograph by the author.

articulation, and these represent cultural spaces that are no less important than the designated places of arts and cultural activity.

*

Conclusion

> While societies and cultures are not the same thing, they are inextricably connected because culture is created and transmitted to others in a society. Cultures are not the product of lone individuals. They are the continuously evolving products of people interacting with each other.
>
> (O'Neil 2006)

To paraphrase Herbert Read on the isolation of the artist: 'an island is only defined by reference to another land mass' (1964, 18), and artworks, artefacts, performances, literature, etc. only become living culture when engaged with the public – eventually and in some way – and this normally takes place in cultural space (even the humble library service). Cultural spaces are also needed to first experience and hone cultural and creative skills, to collaborate and to share with co-participants and publics. Cultural space is therefore only truly *cultural* when experienced, used and valued, although not necessarily through direct or personal experience (i.e. 'non-use value'). Economists use this concept to widen the valuation of cultural assets beyond the utilitarian, where *non-users* are also beneficiaries and still place a value on heritage or the existence of cultural assets, even where they never actually visit or experience them (e.g. for the benefit of others/future generations, the nation/human kind). This *public good* argument only holds good however if there is access for all, and such spaces are non-exclusive – and even when they are notionally accessible, if they are over-crowded or poorly presented, this also diminishes the public good claim, for example due to over-popular exhibitions, *sold out* productions, financial/cost crowding out, or limited programming.

Access and rights to culture also require proximity and equitable spatial distribution, as argued throughout this book and earlier (Evans 2001, 2016b). This is only attainable if cultural space and the everyday are consonant, where the everyday is a proxy for not just access and opportunity, but for participation, engagement and cultural development at all levels. Higher scale and 'special' cultural spaces and production (i.e. institutional, designated and *high art* places and events), and everyday culture, should not therefore be considered as separate or hierarchical, but should be viewed as a continuum, with opportunity not just for access at each level, but across these spaces, including local, regional and national scales. A pyramid of opportunity would seek to ensure that local and other levels of provision and activity can be linked to higher scales of facility (e.g. amateur to professional, small- to medium-scale, youth to adult, multi- to inter-cultural). As we have seen, the multi-use and multi-art form centre does offer a more universal facility model, and the expansion of local and larger-scale arts centres in many countries reflects this demand and potential, but this model also needs to be flexible and responsive to changing cultural tastes and forms – including the intercultural, local and marginal(ised) production, as well as new technologies and the availability of ways to access them (Evans 2001, 133).

Furthermore, whilst formal cultural spaces are designed, curated, programmed and controlled largely by intermediaries (rather than by artists or society at large) – conceived for an otherwise uninvolved public until revealed – the elite effectively talks to elite, a self-referential closed shop maintains a distance and perceived mystique in the eyes of prospective cultural consumers. For instance, as Harris argues in the case of museums, the growth of interactive interpretation strategies

> often privilege curatorial pre-determined responses and presuppose that visitors are disembodied, that actual bodies moving through exhibition spaces are not, in fact, the palpable reality of a museum experience . . . by denying the corporeal presence of visitors, museums continually misrecognise their own institutional identity as they theorize themselves as separate from the visitor.
>
> (2015, 1)

Once placed in a communal space (real or virtual), however, culture takes on a life of its own, and only through this experience and process is culture developed and critiqued.

Most cultural production is of course highly prescribed, codified and controlled, and here the early activism of Arts Labs, community arts centres, street art and flexible/open production (including designer-making and creative-digital networks), as the book has highlighted, has provided an antidote to the slow-moving cathedrals of culture (Lorente 1998). When viewed from within, these formal cultural spaces and institutional cultures are also heavily governed by routine, more so than so-called everyday culture and cultural production (e.g. attending a pottery or dance class, amateur dramatics group, crafts-making workshop, band practice, or oral history group), with prescribed opening-closing times, cleaning, maintenance, curating and programming, etc. Exhibitions and productions are also planned at least one or more years ahead, leaving little or no scope for spontaneity, creative responses or currency in the cultural content they offer. The discipline of the professional individual artist or ensemble likewise requires repetition (and rehearsal/practice is only purposeful when the intention is to share in some way), even where the work is based on the 'classics' or the familiar (e.g. in art, music, plays, literature), which are reproduced day after day/night after night in exhibitions, performances and *on tour*. An observation on much contemporary arts and entertainment production is its highly derivative and low risk nature – sequels, reworked classics and revivals, blockbusters, the over-familiar and safe programming, further extended by digital makeovers. So it would be reasonable to ask whether cultural production and consumption possibilities are best served by the current system and whether the cultural spaces that are the vehicles for this are reflective of contemporary cultural aspirations and society as a whole.

So when considering cultural space production and consumption, the very nature of culture and its distinctions needs primary consideration – between the ideas and practice of creativity, heritage, tradition and innovation, because, as this book argues, these are not simply interchangeable and understanding these distinctions demands a more open and co-created/co-produced approach to the selection, valorisation and distribution of cultural resources – not perhaps through the extreme of a zero-base (wiping the arts and cultural portfolio slate clean), but certainly through more comprehensive cultural mapping, planning and democratisation of much cultural life and therefore opportunity (Kunzmann 2004a). By extension, this would also help to generate more representative creative industries and reduce the embedded social divide in the take-up of cultural activity and the perceived high arts/everyday culture dialectic. More democratic approaches to local amenity provision are seen in cities like New York for example,[11] where each local councillor is allocated a participatory budget of $1.5 million to fund proposals for local initiatives and schemes (e.g. local improvements to schools, parks, libraries and other public spaces) which cost at least $50,000 and last five years or more, and which are then honed and the subject of local elections from which final selection is made and then implemented with capital and other budgets attached ($30 million a year) and facilitated through local not-for-profit organisations, not the council itself. This resembles the Spanish model, where local *Casa de la Cultura* and other amenities are not just developed locally, but governed and prioritised at neighbourhood level, a model also being rolled out in cities such as Quebec in Canada. Examples of participatory cultural space engagement presented here offer an approach that can inform and facilitate this decision-making process. This is particularly important for heterogeneous cultural spaces that differ qualitatively from local amenities such as schools, health and recreation (e.g. parks) facilities.

This importance of engagement in place is no more evident than in the nature and *place* of first-hand cultural experience at a young age, as a determinant of subsequent adult interest. The important role of accessible cultural centres in encouraging participation was observed in research into adult participation and attendance, and earlier childhood exposure to the arts, which made a strong, positive linkage between participatory experience in informal settings (e.g. youth, community arts centre) as opposed to formal and passive attendance at school, theatre trip or museum visit (Dobson and West 1988; Morrison and West 1986). This correlation was high, irrespective of economic and educational backgrounds – key determinants of cultural capital. This was echoed in Harland and Kinder's insightful study of attitudes and barriers to youth participation which concluded that there was 'very little evidence on the school's contribution to encouraging applied and independent engagement with cultural venues . . . schools could help turn young people on to the arts, but they could also turn them off' (1999, 36–37). They also recommended that initial engagement should be more entertainment and experiential rather than educationally motivated (i.e. *curriculum*) and that 'change through dramatic conversions experienced at single arts events is less common than sustained support from significant others who mediate the arts over a period of time' (ibid.) – a sentiment shared with Williams and Braden in the 1970s. So cultural space can be anywhere, anytime, not just the prescribed and conceived, with access and inviting but no less exciting community spaces requiring priority, including intangible and everyday heritage, and interaction through social engagement with artists and, where called-for, activists.

Cultural production – as a beneficiary of the arts-culture-creative continuum (and the source of wealth generation from the creative industries) – can also be exclusive in terms of place, economic and symbolic power (Zukin 1995), and this is reflected in new as well as traditional cultural products and services. Despite a certain democratisation of cultural content and dissemination via digital systems, divides and access are also skewed by big tech platforms, advertising and data privacy/copyright models which perversely appear to support the growth in porn, fraud, fake news and impending artificial intelligence (AI), more so than community culture. But production and innovation need space, and workspace availability and cultural skills and facilities are therefore important, particularly through production quarters and districts which have proven to be valuable in sustaining shared creative spaces and knowledge/facilities exchange. This includes open learning alongside experts and developing synergies across disciplines and backgrounds – not dissimilar from the arts centre (and arts in education) model in cultural development. Resistance strategies and more sustainable models of cultural production are however required in order to counter the more deleterious effects of place branding as urban cultural life is allowed to degenerate through the mono-use of space and an imbalance between, on the one hand, consumption and production, and on the other hand, the commodification of everyday space.

If urban space is, as Lefebvre maintained over 50 years ago, socially produced, then cultural spaces might be expected to be culturally produced within a social structure, but their production-creation-consumption properties still portray the spatial formations and hierarchy of perceived, conceived and lived space with the tensions that this relationship and variable experience implies. These divides in cultural space production – from creation, renewal and usage – as in other produced urban space, become apparent when access is lost or denied, behaviour or interpretation is controlled, prescribed or even ignored or not acknowledged. Like the artist Rachel Whiteread's *inside-out* house,[12] the void becomes the object, revealing the space and form from within – in this metaphor, the (anti-social) space becomes concrete and impervious.

Citizens populate and validate, when and where they can (critically, for better or worse) cultural space, but in formal cultural settings, this is not naturally or necessarily positive or sufficient to fulfil their own aspirations and demands. Here, cultural space production has historically been exclusive, and since we experience most of our engagement through cultural activities outside of the institutional sphere (Lancaster 2010), attention to vernacular culture and accessible places of cultural development and experience would not only ensure that cultural rights are better fulfilled but also encourage cultural institutions to consider the benefits of co-produced and lived cultural space (including their design/adaptation and usage), rather than rationalise their existence and public resources through externalities – economic, social benefits and values – and not through culture and cultural value itself (Kaszynska 2017).

Notes

1 www.mosleyart.com/socially-engaged-art.html
2 www.tate.org.uk/art/art-terms/s/socially-engaged-practice
3 Telephone Interview with Sophie Hope, in Hope (2017).
4 In the UK, only 18% of those working in the performing and visual arts are from a working-class background (vs. c. 50% the population) and only 4% from Black, Asian or Minority Ethnic (BAME) background (vs. 20% of the population).
5 Since the 1930s, Scots migrated to Corby to work in the steel mill, representing over 10% of the town's growing population. The town had the largest Rangers FC supporters club and Scottish Orange (Unionist) Parade outside of Glasgow. The steel mill closed in 1980; it had employed 50% of the town's workforce.
6 www.forwallswithtongues.org.uk/artists/dr-loraine-leeson-docklands-poster-project/
7 www.forwallswithtongues.org.uk/projects/the-changing-picture-of-docklands-1985/
8 www.academia.edu/37392569/Community_Futures_and_Utopia_yesterday_today_ and_tomorrow_re_using_strategies_for_Hornsey_Town_Hall
9 https://vimeo.com/134902583
10 www.leevalley.org/cultural-mappinggis-participation.html
11 https://council.nyc.gov/pb/
12 Whitread's work was a concrete cast of the inside of an entire three-story ordinary house (slated for demolition) – basement, ground floor and first floor, including stairs and bay windows. Winning the Turner Prize, *House* became a popular attraction, with thousands of visitors per day, and attracting graffiti on one side reading "Wot for?" with the enigmatic reply "Why not!". Critic Andrew Graham-Dixon of *The Independent* described it as 'one of the most extraordinary and imaginative public sculptures created by an English artist this century'. A petition demanding it remain permanently received 3,300 signatures. A motion for its retention was moved in the House of Commons by the local MP, but following a close vote locally, it was demolished by the incumbent local council after 11 weeks.

References

Aage, T. and Belussi, F. (2008) 'From Fashion to Design: Creative Networks in Industrial Districts', *Industry and Innovation*, 15(5): 475–491.

ACE (2003) *Focus on Cultural Diversity: Attendance, Participation and Attitudes*. London: Arts Council of England.

ACE (2017) *Taking Part 2016/17: Museums & Galleries*. London: Arts Council England.

Addley, E. (2023) 'Museum on Hunt for Bowie's Dress and James Bond Shirts', *The Guardian*, 23 January, 13.

Adorno, T.W. (ed) (1991) *The Culture Industry: Selected Essays on Mass Culture*. London: Routledge.

Adorno, T.W. and Horkheimer, M. (1943) 'The Culture Industry: Enlightenment as Mass Deception', in *Dialectic of Enlightenment* (trans. J. Cumming). New York: Seabury, 29–48.

Aiesha, R. and Evans, G.L. (2017) 'VivaCity: Mixed-use and Urban Tourism', in M. Smith (ed) *Tourism, Culture and Regeneration*. Wallingford: CABI, 35–48.

Alexander, B. and Alvarado, D.O. (2017) 'Convergence of Physical and Virtual Retail Spaces: The Influence of Technology on Consumer In-Store Experience', in A. Vecchi (ed) *Advanced Fashion Technology and Operations Management*. Hershey, PA: IGI Global, 191–219.

Ando, T. and Futagawa, K. (1990) 'Kara-za: A Movable Theater', *Perspecta Theater, Theatricality, and Architecture*, 26: 171–184.

Andra, I. (1987) 'The Dialetic of Tradition and Progress', in G. Stanishev (ed) *Architecture and Society: In Search of Context*. Sofia: Balkan State Publishing, 156–158.

Andres, L. and Gresillon, B. (2014) 'European Capital of Culture: Leverage for Regional Development and Governance? The Case of Marseille Provence 2013', *Regions*, 295: 3–5.

Anholt, S. (2006) *Anholt City Brand Index – How the World Views Its Cities*, 2nd ed. Bellvue, WA: Global Market Insight.

Appleyard, D. (1981) *Livable Streets*. Berkeley, CA: University of California Press.

Arets, W. (2005) *Living Library: University Library Utrecht*. London: Prestel Publishing Ltd.

Arms, S. (2011) 'The Heritage of Berlin Street Art and Graffiti Scene. Art, Inspiration, Legacy', *Smashing Magazine*, July 13, 1–16. www.smashingmagazine.com/2011/07/13/the-heritage-of-berlinstreet-art-and-graffiti-scene.

Armstrong, S. (2019) *Street Art*. London: Thames & Hudson.

Arnold, D. (2016) 'The Architectural Heritage of Cities: Some Thoughts on Research Methods, Theories and Strategies for Preservation and Sustainable Re-use in a Global Context', *The 4th International Symposium on Architecture Heritage Preservation and Sustainable Development*. Tianjin, China: Urban Flux, October, 180–183.

Ashworth, G. and Tunbridge, J. (2001) *The Tourist-Historic City*. London: Routledge.

Ashworth, G.J. (1994) *Let's Sell Our Heritage to Tourists?* London: London Council for Canadian Studies.

Ashworth, G.J. and Voogd, H. (1990) *Selling the City: Marketing Approaches in Public Sector. Urban Planning*. London: Belhaven Press.

ATCM (2009) *Light Night Network*. www.lightnight.co.uk

Avramidis, K. (2012) *Live Your Greece in Myths: Reading the Crisis on Athens' Walls*. Trento: Professional Dreamers.

Backlund, A.-K. and Sandberg, A. (2002) 'New Media Industry Development: Regions: Networks and Hierarchies – Some Policy Implications', *Regional Studies*, 36(1): 87–91.

Bagwell, S., Evans, G., Witting, A. and Worpole, K. (2012) *Public Space Management: Report to the Intercultural Cities Programme*. Strasbourg: Council of Europe.

Bakhshi, H., Frey, C. and Osborne, M. (2015) *Creativity vs Robots. The Creative Economy and the Future of Employment*. London: NESTA.

Bakhshi, H., McVittie, E. and Simmie, J. (2008) *Creating Innovation: Do the Creative Industries Support Innovation in the Wider Economy?* London: Experian.

Banks, M. (2023) 'Cultural Work and Contributive Justice', *Journal of Cultural Economy*, 16(1): 47–61.

Banksy (2002) *Existencillism*. London: Weapons of Mass Distraction.

Basu, P. (2013) 'Memoryscapes and Multi-Sited Methods', in E. Keightley and M. Pickering (eds) *Research Methods for Memory Studies*. Edinburgh: Edinburgh University Press, 115–131.

Bauer, O. (2000) *The Nationalities Question and Social Democracy*. Minneapolis: University of Minnesota Press.

Bauman, Z. (1996) 'From Pilgrim to Tourist – Or a Short History of Identity', in S. Hall and P. Du Gay (eds) *Questions of Cultural Identity*. London: Sage, 18–36.

Bazelman, J. (2014) *The Valuation of Cultural Heritage: A Roadmap*. Amsterdam: Dutch Heritage Agency.

Beauvert, T. (1995) *Opera Houses of the World*. New York: The Vendome Press.

Beech, D. (2004) *I Fail to Agree: Hewitt and Jordan*. Sheffield: Site Gallery.

Belfiore, E. (2022) 'Who Cares? At What Price? The Hidden Costs of Socially Engaged Arts Labour and the Moral Failure of Cultural Policy', *European Journal of Cultural Policy*, 25(1): 671–678.

Bell, C. and Newby, H. (1976) 'Community, Communion, Class and Community Action', in D. Herbert and R. Johnson (eds) *Social Areas in Cities*. London: Wiley.

Bell, D. and Jayne, M. (eds) (2004) *City of Quarters. Urban Villages in the Contemporary City*. Aldershot: Ashgate.

Benigson, M. (2014) *Streetwear and Couture: Blurring Lines in the Fashion Industry*. London: The MSB Group.

Benjamin, W. (1935/1979) 'The Work of Art in the Age of Mechanical Reproduction', in *Illuminations* (trans. H. Zohn). London: Fontana, 219–253.

Benjamin, W. (1999) *Selected Writings, Volume 2: 1927–1934*. Cambridge, MA: Belknap/Harvard University Press.

Bennett, T. (1988) 'The Exhibitionary Complex', *New Formations*, 4: 73–104.

Bianchini, F. and Parkinson, M. (eds) (1993) *Cultural Policy and Urban Regeneration: The West European Experience*. Manchester: Manchester University Press.

Birdsall, C. (2013) '(In)audible Frequencies: Sounding Out the Contemporary Branded City', in C. Lindner and H. Hussey (eds) *Paris-Amsterdam Underground*. Amsterdam: Amsterdam University Press, 115–131.

Bishop, C. (2006) 'The Social Turn: Collaboration and Its Discontents', *Artforum*, February, 178–183.

Black, L. (2006) '"Making Britain a Gayer and More Cultivated Country": Wilson, Lee and the Creative Industries in the 1960s', *Contemporary British History*, 20(3): 323–342.

Boddy, T. (2006) 'The Library and the City', *Architectural Review*, 44: 1–46.

Boland, P., Murtagh, B. and Shirlow, P. (2019) 'Fashioning a City of Culture: "Life and Place Changing" or a "12-Month Party"', *International Journal of Cultural Policy*, 25(2): 246–265.

Bolognesi, C. (2005) *Milan and Lombardy: The Revival of the Future*. Milan: Lombardy Region/Milan City.

Bone, J., et al. (2021) 'Who Engages in the Arts in the United States? A Comparison of Several Types of Engagement using Data from the General Social Survey,' *BMC Public Health*, 21(1349): 1–45.

BOP (2016) *East London Fashion Cluster*. London: BOP.

Bordage, F. (2002) *TransEuropeHalles: The Factories Conversions for Urban Culture*. Basel: Birkhauser.

Borghini, S., Visconti, L.M., Anderson, L. and Sherry, J. (2010) 'Symbiotic Postures of Commercial Advertising and Street Art', *Journal of Advertising*, 39(3): 113–126.

Boschma, R. (2005) 'Proximity and Innovation: A Critical Assessment', *Regional Studies*, 39: 61–74.

Bourdieu, P. (1983/1993) *The Field of Cultural Production* (trans. R. Johnson). Cambridge: Polity Press.

Bourdieu, P. (1984) *Distinction: A Social Critique of the Judgment of Taste*. Cambridge, MA: Harvard University Press.

Bourdieu, P. and Darbel, A. (1969/1991) *The Love of Art*. Cambridge: Polity Press.

Bovone, L. (2005) 'Fashionable Quarters in the Postindustrial City: The Ticinese of Milan', *City and Community*, 4(4): 359–380.

Boyko, C.T. (2003) 'Breathing New Life into Old Places Through Culture: A Case of Bad Breath?', in G. Richards and J. Wilson (eds) *From Cultural Tourism to Creative Tourism: Changing Places, the Spatial Challenge of Creativity (Part 3)*. Arnhem: ATLAS, 19–31.

Braddock Clarke, S.E. and Harris, J. (2012) *Digital Visions for Fashion and Textiles: Made in Code*. London: Thames & Hudson.

Braden, S. (1977) *Artists and People*. London: Routledge & Kegan Paul.

Breheny, M. (1996) 'Centrists, Decentrists and Compromisers: Views on the Future of Urban Form', in M. Jenks, E. Burton and K. Williams (eds) *The Compact City: A Sustainable Urban Form?* London: Pion.

Breward, C. (2004) *Fashioning London. Clothes and the Modern Metropolis*. Oxford: Berg.

Breward, C. and Gilbert, D. (eds) (2006) *Fashion's World Cities*. Oxford: Berg.

Breward, C. and Gilbert, D. (2010) *Fashion Cities. Berg Encyclopaedia of World Dress and Fashion*. London: Bloomsbury.

Breznitz, S. and Noonan, D. (2014) 'Arts Districts, Universities, and the Rise of Digital Media', *The Journal of Technology Transfer*, 39(4): 594–615.

Brighenti, M. (2010) 'At the Wall: Graffiti Writers, Urban Territoriality, and the Public Domain', *Space and Culture*, 13(3): 315–332.

Brook, O. (2011) *International Comparisons of Public Engagement in Culture and Sport*. London: DCMS.

Brook, O., Boyle, P. and Flowerdew, R. (2010) 'Geographic Analysis of Cultural Consumption', in J. Stillwell, P. Norman, C. Thomas and P. Surridge (eds) *Spatial and Social Disparities: Understanding Population Trends and Processes – Volume 2*. Dordrecht: Springer, 67–82.

Brook, O., O'Brien, D. and Taylor, M. (2020) *Culture Is Bad for You: Inequality in the Cultural and Creative Industries*. Manchester: Manchester University Press.

Brooks, L. (2022) 'The People's Gallery. Legal Walls Allow Street Art to Thrive in Glasgow', *The Guardian*, 6 January, 7.

Brown, M. (2014) 'Gormley and Company Send Art All Over the Place', *The Guardian*, 17 July, 11.

Brown, W. (2020) *A New Way to Understand the City: Henri Lefebvre's Spatial Triad*. https://will-brown.medium.com/a-new-way-to-understand-the-city-henri-lefebvres-spatial-triad-d8f800a9ec1d (accessed 29 December 2022).

Burgers, J. (1995) 'Public Space in the Post-Industrial City', in G. Ashworth and A. Dietvorst (eds) *Tourism and Spatial Transformations: Implications for Policy and Planning*. Wallingford: CAB International, 147–161.

Burgess, J.A. (1982) 'Selling Places: Environmental Images for the Executive', *Regional Studies*, 16: 1–17.

Burtenshaw, D., Bateman, M. and Ashworth, G.J. (1991) *The European City: A Western Perspective*. London: David Fulton.

Butler, R.W. (1980) 'The Concept of the Tourism Area Life Cycle of Evolution: Implications for Management of Resources', *Canadian Geographer*, 24(1): 5–12.

Cairncross, F. (1995) 'The Death of Distance: A Survey of Telecommunications', *The Economist*, 336(7934): 5–28.

Calvino, I. (1979) *Invisible Cities*. London: Pan.

Carbon, C.C. (2017) 'Art Perception in the Museum: How We Spend Time and Space in Art Exhibitions', *Iperception*, 18(1): 2041669517694184. https://doi.org/10.1177/2041669517694184.

Carter-Morley, J. (2021) 'Fashion Blockbuster Locations Showcase Return of Dressing Up', *The Guardian*, 19 June, 21.

Castells, M. (1991) *The Informational City Information Technology, Economic Restructuring, and the Urban-Regional Process*. Oxford: Blackwell.

Castells, M. (1996) *The Rise of the Network Society*. Oxford: Blackwell.

Castells, M. (1997) 'Citizen Movements, Information and Analysis: An Interview with Manuel Castells', *City*, 7: 140–155.

Catungal, J.P., Leslie, D. and Hii, Y. (2009) 'Geographies of Displacement in the Creative City: The Case of Liberty Village, Toronto', *Urban Studies*, 46(5/6): 1095–1114.

Cauchi-Santoro, R. (2016) 'Mapping Community Identity: Safeguarding the Memories of a City's Downtown Core', *City, Culture and Society*, 7(1): 43–54.

CCV (2022) *Research Digest: Everyday Creativity*, Vol. 1. Leeds: Centre for Cultural Value.

CEC (1992) *European Ministers of Culture Meeting Within the Council on Guidelines for Community Cultural Action*. Brussels: Official Journal of the European Communities Council (CEC).

Chanan, M. (1980) *The Dream that Kicks: The Prehistory and Early Years of Cinema in Britain*. London: Routledge & Kegan Paul.

Chang, V. (2013) 'Animating the City: Street Art, Blu and the Poetics of Visual Encounter', *Animation*, 8(3): 215–233.

Chapman, J. (2011) *Emotionally Durable Design*. London: Earthscan.

Charnock, G. and Ribera-Fumaz, R. (2014) 'The Production of Urban Competitiveness: Modelling22@Barcelona', in L. Stanek, C. Schmid and A. Moravanszky (eds) *Urban Revolution Now. Henri Lefebvre in Social Research and Architecture*. Farnham: Ashgate, 157–171.

Chen, J., Judd, B. and Hawken, S. (2015) *Adaptive Reuse of Industrial Heritage for Cultural Purposes in Three Chinese Mega-Cities: Beijing, Shanghai and Chongqing*. Sydney: RICS/COBRA AUBEA.

Cheshire, J. and Uberti, O. (2016) *London: Information Capital*. London: Penguin.

Chilese, E. and Russo, A.P. (2008) *Urban Fashion Policies: Lessons from the Barcelona Catwalks, EBLA Working Paper 200803*. Turin: University of Turin.

Christophers, B. (2008) 'The BBC, the Creative Class, and Neoliberal Urbanism in the North of England', *Environment and Planning A: Economy and Space*, 40(10): 2313–2329.

Çinar, A. and Bender, T. (eds) (2007) *Urban Imaginaries: Locating the Modern City*. Minnesota: University of Minnesota Press.

Cinderby, C. and Evans, G.L. (2013) *Proceedings of INCLUDE Asia*. London: Royal College of Art, 68–76.

Cities Institute (2010) 'Mapping the Digital Economy', *Digital Shoreditch*, May, 1–13.

City of Copenhagen (2003) *Orestad. Historic Perspective, Planning, Implementation, Documentation*. Copnhagen: City of Copenhagen.

City of Toronto (2010) *Creative City Planning Framework: A Supporting Document to the Agenda for Prosperity: Prospectus for a Great City*. Toronto: Authenticity/City of Toronto.

Cohen, E. (1999) 'Cultural Fusion', in *Values and Heritage Conservation*. Los Angeles: Getty Conservation Institute, 44–50.

Cohen, P. (2013) *On the Wrong Side of the Track? East London and the Post-Olympics*. London: Lawrence & Wishart.

Colliers International (2011) *Encouraging Investment in Heritage at Risk*. London: English Heritage.

Connor, C. and Barlow, G. (2023) 'Could Arts Centres Hold the Key to UK Culture's Future?', *The Guardian*, 3 April. www.theguardian.com/culture-professionals-network/culture-professionals-blog/2014/apr/03/art-centres-uk-culture-future (accessed 22 March 2023).

Council of Europe (1992) *European Urban Charter*. Strasbourg: Standing Conference of Local and Regional Authorities of Europe (CLRAE).

Council of Europe (2008) *Living Together as Equals in Dignity. White Paper on Intercultural Dialogue*. Strasbourg: Council of Europe.

Craik, J. (1997) 'The Culture of Tourism', in C. Rojek and J. Urry (eds) *Touring Cultures: Transformations of Travel and Theory*. London: Routledge, 113–136.

Crane, D. (2000) *Fashion and Its Social Agendas. Class, Gender, and Identity in Clothing*. Chicago and London: The University of Chicago Press.

Creigh-Tyte, S. and Selwood, S. (1998) 'Museums in the UK: Some Evidence on Scale and Activities', *Journal of Cultural Economics*, 22(2/3): 151–165.

Creswell, T. (1992) 'The Crucial "Where" of Graffiti: A Geographical Analysis of Reactions to Graffiti in New York', *Environment & Planning D: Society and Space*, 10(3): 329–344.

Creswell, T. (1996) *In Place/Out of Place: Geography, Ideology and Transgression*. Minneapolis: University of Minnesota Press.

Crewe, L. and Beaverstock, J. (1998) 'Fashioning the City: Cultures of Consumption in Contemporary Urban Spaces', *Geoforum*, 29(3): 287–308.

Cronin, A.M. (2008) 'Urban Space and Entrepreneurial Property Relations: Resistance and the Vernacular of Outdoor Advertising and Graffiti', in A.M. Cronin and K. Hetherington (eds) *Consuming the Entrepreneurial City: Image Memory, Spectacle*. New York: Routledge, 65–84.

Crowhurst, A.J. (1992) *The Music Hall, 1885–1922. The Emergence of a National Entertainment Industry in Britain*. Unpublished PhD thesis. Cambridge: University of Cambridge.

Culture24 (2009) *5th Edition of the European Night of Museums*. www.culture24.org.uk

Cummings, N. and Lewandowska, M. (2000) *The Value of Things*. Basel: Birkhauser.

CURDS (2009) *Literature Review: Historic Environment, Sense of Place, and Social Capital*. London: English Heritage.

Curtis, D. (2020) *London's Arts Labs and the 60s Avant-Garde*. London: John Libbey Publishing.

Cuthbert, A. (2006) *The Form of Cities*. Oxford: Blackwell.

Daniels, M. (2013) *Paris National and International Exhibitions from 1798 to 1900: A Finding-List of British Library Holdings*. London: British Library.

Davidson, J. (2022) 'Dub, Utopia and the Ruins of the Caribbean', *Theory, Culture and Society*, 39(1): 3–22.

Davies, R. (2003) *A Map of Toronto's Cultural Facilities: A Cultural Facilities Analysis*. Toronto: City of Toronto Division of Economic Development, Culture and Tourism.

Dawson, R. (2021) *London Street Art and Graffiti Through the Decades: 1960–2021*, May 7. https://blog.bookanartist.co/london-street-art-and-graffiti-through-the-decades-1960-to-2021 (accessed 20 February 2023).

DCMS (2001) *Creative Industries Mapping Document 2001*, 2nd ed. London: Department of Culture, Media and Sport.

DCMS (2010) *Culture and Sport Physical Asset Mapping Toolkit*, Cities Institute and TBR for ACE, Historic England & Sport England. London: DCMS.

DCMS (2014) *Independent Library Report*. London: Department for Culture Media and Sport.

DCMS (2016) *The Culture White Paper*. London: DCMS.

DCMS (2020) *Heritage – Taking Part Survey 2019/20*. London: DCMS.

DCMS (2022) *Scoping Culture and Heritage Capital Report*. London: DCMS.

de Abreu Santos, V.Á. and van der Borg, J. (2023) 'Cultural Mapping Tools and Co-Design Process: A Content Analysis to Layering Perspectives on the Creative Production of Space', *Sustainability*, 15(6): 5335.

Deleuze, G. and Guattari, F. (1987) *A Thousand Plateaus: Capitalism and Schizophrenia*. London: Continuum.

Delrieu, V. and Gibson, L. (2017) 'Libraries and the Geography of Use: How Does Geography and Asset "Attractiveness" Influence the Local Dimensions of Cultural Participation?', *Cultural Trends*, 26: 18–33.

DeNotto, M. (2014) 'Street art and Graffiti. Resources for Online Study', *C&RL News*, April, 208–211.

De Propris, L. and Hypponen, L. (2008) 'Creative Clusters and Governance: The Dominance of the Hollywood Film Cluster', in P. Cooke and L. Lazzeretti (eds) *Creative Cities, Cultural Clusters and Local Development*. Cheltenham: Edward Elgar, 340–371.

DeSilvey, C., Blundell, A., Fredheim, H. and Harrison, R. (2022) *Identifying Opportunities for Integrated Adaptive Management of Heritage Change and Transformation in England: A Review of Relevant Policy and Current Practice*. Research Report 18/22. London: Historic England.

Dinnie, K. (2004) 'Place Branding: Overview of an Expanding Literature', *Place Branding and Pubic Diplomacy*, 1(1): 106–110.

Di Vita, S. (2020) 'The Milan EXPO 2015', in G.L. Evans (ed) *Mega-Events, Placemaking, Regeneration and City-Regional Development*. London: Routledge, 70–86.

Dobson, L.C. and West, E.G. (1988) 'Performing Arts Subsidies and Future Generations', *Journal of Cultural Economics*, 12: 8–115.

Dodd, F. (2008) *Our Creative Talent: The Voluntary and Amateur Arts in England*. London: DCMS.

Doeringer, P.B. and Crean, S. (2006) 'Can Fast Fashion Save the us Apparel Industry?', *Socio-Economic Review*, 4(3): 353–377.

Downey, J. and McGuigan, J. (eds) (1999) *Technocities*. London: Sage.

Duncum, P. (2002) 'Theorising Everyday Aesthetic Experience with Contemporary Visual Culture', *Visual Arts Research*, 28(2): 4–15.

Duxbury, N. (2004) *Creative Cities: Principles and Practices*. Background Paper F47. Ottawa: Canadian Policy Research Networks Inc.

Duxbury, N., Garett-Petts, W. and MacLennan, D. (eds) (2015) *Cultural Mapping as Cultural Inquiry*. London: Routledge.

Duxbury, N., Garrett-Petts, W. and Longley, A. (eds) (2018) *Artistic Approaches to Cultural Mapping. Activating Imaginaries and Means of Knowing*. London: Routledge.

Duxbury, N. and Redaelli, E. (2020) 'Cultural Mapping', in P. Moy (ed) *Oxford Bibliographies in Communication*. New York: Oxford University Press.

Ebrey, J. (2016) 'The Mundane and Insignificant, the Ordinary and the Extraordinary: Understanding Everyday Participation and Theories of Everyday Life', *Cultural Trends*, 25(3): 158–168.

EC (2009) *Preserving Our Heritage, Improving Our Environment, Volume I. 20 Years of EU Research into Cultural Heritage*. Brussels: European Commission (EU) DG Research.

Edensor, T. (1998) *Tourists at the Taj: Performance and Meaning at a Symbolic Site*. London: Routledge.

Edizel-Tasci, O. and Evans, G.L. (2017) 'Participatory Mapping and Engagement with Urban Water Communities', in A. Ersoy (ed) *The Impact of Co-production*. Bristol: Policy Press, 119–136.

Edizel-Tasci, O. and Evans, G.L. (2021) 'Community Engagement in Climate Change Policy: The Case of Three Mills, East London', in E. Peker and A. Ataöv (eds) *Governance of Climate Responsive Cities*. Vienna: Springer.

Edizel-Tasci, O., Evans, G.L. and Dong, H. (2013) 'Dressing up London', in V. Girginov (ed) *Handbook of the London 2012 Olympic and Paralympic Games*, Vol. 2. London: Routledge, 19–35.

EH (2000) *Power of Place: The Future of the Historic Environment*. London: English Heritage.

EH (2003) *Heritage Counts*. London: English Heritage.

Ehrman, E.W. (2018) *Fashioned from Nature*. London: V&A.

Eicher, J.B. (ed) (2010) *Berg Encyclopaedia of Fashion Cities*. Oxford: Berg.

Elden, S. (2004) *Understanding Henri Lefebvre: Theory and the Possible*. New York: Continuum Books.

Emerling, S. (2001) 'Prada Enters a New Frontier of Retailing', *Los Angeles Times*, 16 April. www.latimes.com/archives/la-xpm-2001-apr-16-cl-51533-story.html (accessed 20 April 2023).

Erdogan, G. (2014) 'Mapping Street Art in the Case of Turkey, Istanbul, Beyoglu Yuksek Kaldirim', *Paper to Lisbon Street Art & Urban Creativity International Conference*. Lisbon: Lisbon University, 3–5 July.

Evans, G.L. (1995) 'Planning for the British Millennium Festival: Establishing the Visitor Baseline and a Framework for Forecasting', *Festival Management and Event Tourism*, 3(4): 183–196.

Evans, G.L. (1998a) 'In Search of the Cultural Tourist and the Post-Modern Grand Tour', *International Sociological Association XIV Congress (RC50)*. Montreal, July.

Evans, G.L. (1998b) 'Urban Leisure: Edge City and the New Leisure Periphery', in M. Collins and I. Cooper (eds) *Leisure Management – Issues and Application*. Wallingford: CAB International, 113–138.

Evans, G.L. (1999a) 'Networking for Growth and Digital Business', in W. Schertler et al. (eds) *ICT in SMEs*. Vienna: Springer-Verlag, 376–387.

Evans, G.L. (1999b) 'The Economics of the National Performing Arts – Exploiting Consumer Surplus and Willingness-to-Pay: A Case of Cultural Policy Failure?', *Leisure Studies*, 18: 97–118.

Evans, G.L. (2000) 'Contemporary Crafts as Artefacts and Functional Goods and Their Role in Local Economic Diversification and Cultural Development', in M. Hitchcock and K. Teague (eds) *Souvenirs: The Material Culture of Tourism*. Aldershot: Ashgate, 127–146.

Evans, G.L. (2001) *Cultural Planning: An Urban Renaissance?* London: Routledge.

Evans, G.L. (2002) 'Living in a World Heritage City: Stakeholders in the Dialectic of the Universal and the Particular', *International Journal of Heritage Studies*, 8(2): 117–135.

Evans, G.L. (2003a) 'Hard Branding the Culture City – From Prado to Prada', *International Journal of Urban and Regional Research*, 27(2): 417–440.

Evans, G.L. (2003b) 'Whose Heritage Is It Anyway? Reconciling the "National" and the Universal in Quebec City', *British Journal of Canadian Studies*, 16(2): 343.

Evans, G.L. (2004) 'Cultural Industry Quarters: From Pre-Industrial to Post-Industrial Production', in D. Bell and M. Jayne (eds) *City of Quarters. Urban Villages in the Contemporary City*. Aldershot: Ashgate, 71–92.

Evans, G.L. (2005) 'Measure for Measure: Evaluating the Evidence of Culture's Contribution to Regeneration', *Urban Studies*, 42(5/6): 959–983.

Evans, G.L. (2006) 'Branding the City: The Death of City Planning?', in J. Monclus (ed) *Culture, Urbanism & Planning*. Aldershot: Ashgate, 197–214.

Evans, G.L. (2007) 'Tourism, Creativity and the City', in G. Richards and G.J. Wilson (eds) *Tourism Creativity & Development*. London: Routledge, 35–48.

Evans, G.L. (2008) 'Cultural Mapping and Sustainable Communities: Planning for the Arts Revisited', *Cultural Trends*, 17(2): 65–96.

Evans, G.L. (2009a) 'Urban Sustainability: Mixed Use or Mixed Messages?', in C. Boyko, R. Cooper and G.L. Evans (eds) *Designing Sustainable Cities*. Oxford: Wiley-Blackwell, 190–217.

Evans, G.L. (2009b) 'From Cultural Quarters to Creative Clusters: Creative Spaces in the New City Economy', in M. Legner (ed) *The Sustainability and Development of Cultural Quarters: International Perspectives*. Stockholm: Institute of Urban History, 32–59.

Evans, G.L. (2009c) 'Creative Cities, Creative Spaces and Urban Policy', *Urban Studies*, 46(5&6): 1003–1040.

Evans, G.L. (2009d) 'Creative Spaces and the Art of Urban Living', in T. Edensor, D. Leslie, S. Millington and N. Rantisi (eds) *Spaces of Vernacular Creativity: Rethinking the Cultural Economy*. London: Routledge, 19–32.

Evans, G.L. (2010) 'Heritage Cities', in R. Beauregard (ed) *Encyclopaedia of Urban Studies*. New York: Sage, 136–138.

Evans, G.L. (2011) 'Cities of Culture and the Regeneration Game', *London Journal of Tourism, Sport and Creative Industries*, 5(67): 5–18.

Evans, G.L. (2012a) 'Hold Back the Night: Nuit Blanche and All-Night Events in Capital Cities', *Current Issues in Tourism*, 15(1–2): 35–49.

Evans, G.L. (2012b) 'Creative Small and Medium-Sized Cities', *International Journal of Cultural Administration*, 1: 141–157.

Evans, G.L. (2013a) 'Cultural Planning and Sustainable Development', in G. Baker and D. Stevenson (eds) *The Ashgate Research Companion to Planning and Culture*. London: Routledge, 223–228.

Evans, G.L. (2013b) 'Maastricht: From Treaty Town to European Capital of Culture', in C. Grodach and D. Silver (eds) *The Politics of Urban Cultural Policy: Global Perspectives*. London: Routledge, 264–285.

Evans, G.L. (2014a) 'Rethinking Place Branding and Place Making Through Creative and Cultural Quarters', in M. Kavaratzis, G. Warnaby and G. Ashworth (eds) *Rethinking Place Branding: Comprehensive Brand Development for Cities and Regions*. Vienna: Springer, 135–158.

Evans, G.L. (2014b) 'Living in the City Mixed Use and Quality of Life', in R. Cooper, E. Burton and C. Cooper (eds) *Wellbeing and the Environment: Wellbeing: A Complete Reference Guide*, Vol. II. Oxford: John Wiley.

Evans, G.L. (2014c) 'Accessibility and User Needs: Pedestrian Mobility and Urban Design in the UK', *Municipal Engineer*, 168(1): 1–13.

Evans, G.L. (2015a) 'Cultural Mapping and Planning for Sustainable Communities', in N. Duxbury, W. Garett-Petts and D. MacLennan (eds) *Cultural Mapping as Cultural Inquiry*. London: Routledge, 45–68.

Evans, G.L. (2015b) 'Designing Legacy and the Legacy of Design: London 2012 and the Regeneration Games', *Architectural Review Quarterly*, 18(4): 353–366.

Evans, G.L. (2016a) 'London 2012', in J. Gold and M. Gold (eds) *Olympic Cities. City Agendas, Planning and the Worlds Games, 1896–2016*. London: Routledge.

Evans, G.L. (2016b) 'Participation and Provision in Arts & Culture – Bridging the Divide', *Cultural Trends*, 25(1): 2–20.

Evans, G.L. (2016c) *Place Branding and Heritage, with TBR and Pomegranite Seeds*. London: Historic England Heritage Counts.

Evans, G.L. (2016d) 'Graffiti Art and the City. From Piece-Making to Place-Making', in I. Ross (ed) *Routledge Handbook of Graffiti and Street Art*. London: Routledge, 164–178.

Evans, G.L. (2017) 'Creative Cities: An International Perspective', in G. Richards and J. Hannigan (eds) *The SAGE Handbook of New Urban Studies*. New York: Sage, 311–329.

Evans, G.L. (2018a) 'Designing Contemporary Mega Events', in A. Massey (ed) *Blackwell Companion to Contemporary Design*. Oxford: Blackwell.

Evans, G.L. (2018b) 'Inclusive and Sustainable Design in the Built Environment: Regulation or Human-Centres?', *Built Environment*, 44(1): 79–93.

Evans, G.L. (2018c) *Smart Cities and Waste Innovation*. Global Fashion Conference. http://gfc-conference.eu/wp-content/uploads/2018/12/EVANS_SmART-Cities-and-Waste-Innovation.pdf

Evans, G.L. (2019) 'Emergence of a Digital Cluster in East London: Birth of a New Hybrid Firm', *Competitiveness Review*, 29(3): 253–266.

Evans, G.L. (ed) (2020a) *Mega-Events: Placemaking, Regeneration and City-Regional Development*. London: Routledge.

Evans, G.L. (2020b) 'Events, Cities and the Night-Time Economy', in S. Page and J. Connell (eds) *Routledge Handbook of Events*. London: Routledge.

Evans, G.L. (2020c) 'From Albertopolis to Olympicopolis', in G.L. Evans (ed) *Mega-Events. Placemaking, Regeneration and City-Regional Development*. London: Routledge, 35–52.

Evans, G.L. (2022) *Maximising and Measuring the Value of Heritage in Place*. Paper in 5 in AHRC Future Trends Series. Warwick: Warwick University City of Culture Project.

Evans, G.L., Aiesha, R., and Foord, J. (2009) 'Mixed Use or Mixed Messages?', In R. Cooper, G.L. Evans and C. Boyko (eds) *Designing Sustainable Cities*. Oxford: Blackwell, 190–217.

Evans, G.L. and Foord, J. (2000) 'European Funding of Culture: Promoting European Culture or Regional Growth', *Cultural Trends*, 36: 53–87.

Evans, G.L. and Foord, J. (2006a) 'Rich Mix Cities: From Multicultural Experience to Cosmopolitan Engagement', *Ethnologia Europaea: Journal of European Ethnology*, 34(2): 71–84.

Evans, G.L. and Foord, J. (2006b) 'Small Cities for a Small Country: Sustaining the Cultural Renaissance?', in D. Bell and M. Jayne (eds) *Small Cities, Urban Experience Beyond the Metropolis*. London: Routledge.

Evans, G.L. and House, N. (2017) 'Architecture, Intervention and Adaptive ReUse', *Int/AR Journal*, 8: 26–33.

Evans, G.L., NEF and TBR (2016) *The Role of Culture, Sport and Heritage in Place Shaping*. Culture Evidence (CASE) programme. London: DCMS/ACE/Historic England, Sport England.

Evans, G.L. and Reay, D. (1996) *Arts Culture and Entertainment Park Plan – Topic Study*. Waltham Abbey: Lee Valley Regional Park Authority.

Evans, G.L. and Shaw, S. (2001) 'Urban Leisure and Transport: Regeneration Effects', *Journal of Retail Leisure Property*, 1: 350–372.

Evans, G.L. and TBR (2010) *The Art of the Possible: A Feasibility Study on Assessing the Impact of Cultural and Sporting Investment*. DCMS CASE programme. London: DCMS/ACE/Historic England, Sport England.

Evans, G.L. and Witting, A. (2006) *Creative Spaces, Strategies for Creative Cities: Berlin*. London: LDA Creative.

Fairlie, R.W., London, R.A., Rosner, R. and Pastor, M. (2006) *Crossing the Divide: Immigrant Youth and Digital Disparity in California*. Santa Cruz: University of California, Center for Justice, Tolerance and Community.

Ferrell, J. (1993) *Crimes of Style: Urban Graffiti and the Politics of Criminality*. New York: Garland.

Filion, P. (2019) 'Lefebvre and Contemporary Urbanism. The Enduring Influence and Critical Power of His Writing on Cities', in M. Leary-Owhin and J. McCarthy (eds) *The Routledge Handbook of Henri Lefebvre, The City and Urban Society*. London: Routledge.

Fish, R. and Church, A. (2013) *A Conceptual Framework for Cultural Ecosystem Services Working Paper*. Exeter: Centre for Rural Policy Research: University of Exeter.

Fisher, R. (1993) *The Challenge for the Arts: Reflection on British Culture in Europe in the Context of the Single Market and Maastricht*. London: Arts Council of Great Britain.

Fleming, T. (2010) *North Northants Mapping Overview of Cultural Assets*. London: Tom Fleming Consultancy.

Florida, R. (2002) *The Rise of the Creative Class: And How It's Transforming Work, Leisure, Community and Everyday Life*. New York: Basic Books.

Florida, R. (2003) 'Cities and the Creative Class', *City & Community*, 2(1): 3–19.

Florida, R. (2005) *Cities and the Creative Class*. New York: Routledge.

Foord, J. (1999) 'Creative Hackney: Reflections on Hidden Art', *Rising East*, 3: 69–94.

Foord, J. (2009) 'Strategies for Creative Industries: An International Review', *Creative Industries Journal*, 1(2): 91–113.

Foord, J. (2010) 'Mixed-Use Trade-Offs: How to Live and Work in a "Compact City" Neighbourhood', *Built Environment*, 36(1): 47–62.

Foord, J. (2013) 'The New Boomtown? Creative City to Tech City in East London', *Cities*, 33: 51–60.

Forty, A. (2000) *Words and Buildings: A Vocabulary of Modern Architecture*. London: Thames and Hudson.

Foster, H. (2013) *The Art-Architecture Complex*. London: Verso.

Fraser, A. (2005a) 'From the Critique of Institutions to an Institution of Critique', *Artforum*, 44(1): 283.

Fraser, A. (2005b) '"Isn't This a Wonderful Place?" A Tour of a Tour of the Guggenheim Bilbao, Museum Highlights', in A. Alberro (ed) *The Writings of Andrea Fraser*. Cambridge, MA: MIT Press, 233–260.

Fuller, H., Helbrecht, I., Schlueter, S., Mackrodt, U., Genz, C., Walthall, B., van Gielle Ruppe, P. and Dirksmeier, P. (2018) 'Manufacturing Marginality. (Un-)Governing the Night in Berlin', *Geoforum*, 94: 24–32.

Galloway, S. and Dunlop, S. (2007) 'A Critique of Definitions of the "Cultural and Creative Industries" in Public Policy', *International Journal of Cultural Policy*, 13(1): 17–31.

Garcia, B. (2017) 'Cultural Olympiads', in J. Gold and M. Gold (eds) *Olympic Cities. City Agendas, Planning and the World's Games, 1896–2020*. London: Routledge.

Garcia, B. and Cox, T. (2013) *European Capitals of Culture: Success Strategies and Long-Term Effects*, IP/B/CULT/IC/2012-082. Brussels: European Parliament.

Garnham, N. (1984) *Cultural Industries: What Are They? Views*. London: Independent Film and Video Producers Association.

GAU (Galeria De Arte Urbana) (2014) Vol. 3. www.facebook.com/galeriadearteurbana

Geertz, C. (1985) *Local Knowledge: Further Essays in Interpretive Anthropology*. New York: Basic Books.

Gehl, J. (2001) *Life Between Buildings*. Copenhagen: Danish Architectural Press.

Gehry, F. (2014) 'Editorial', *Wallpaper Magazine*, October. London, 255–272.

Gertler, M. et al. (2006) *Creative Spaces, Strategies for Creative Cities: Toronto*. London and Toronto: LDA Creative London/City of Toronto.

Getty (1999) *Economics and Heritage Conservation*. Los Angeles: Getty Conservation Institute.

Getz, D. (2012) *Event Studies: Theory, Research and Policy for Planned Events*. London: Routledge.

Girard, A. (2001) 'Maison de la culture', in E. de Waresquiel (ed) *Dictionary of Cultural Policies in France Since 1959*. Paris: Larousse/CNRS Editions.

Glendinning, M. (2010) *Architecture Empire? The Triumph and Tragedy of Global Modernism*. London: Reaktion Books.

Gluckman, R. (2007) 'Qatar's Desert Bloom', *Urban Land*, 66(4): 90–93.

Gold, J. and Gold, M. (2016) 'Olympic Futures and Urban Imaginings: From Albertopolis to Olympicopolis', in G. Richards and J. Hannigan (eds) *The Sage Handbook of New Urban Studies*. New York: Sage, 514–534.

Gold, J. and Gold, M. (2020) *Festival Cities: Culture, Planning and Urban Life*. London: Routledge.

Gong, H. and Hassink, R. (2017) 'Exploring the Clustering of Creative Industries', *European Planning Studies*, 25(4): 583–600.

Gonzales, A.M. (2010) 'On Fashion and Fashion Discourses', *Critical Studies in Fashion and Beauty*, 1(1): 65–85.

Gonzales, S. (2006) 'Scalar Narratives in Bilbao: A Cultural Politics of Scales Approach to the Study of Urban Policy', *International Journal of Urban and Regional Research*, 30(4): 836–857.

Gordon, C. (1991) 'Governmental Rationality: An Introduction', in G. Burchill, C. Gordon and P. Miller (eds) *The Foucault Effect: Studies in Governmentality*. London: Harvester Wheatsheaf, 1–51.

Grabow, B., Henckel, D. and Hollbach-Grömig, B. (1995) *Weiche Standortfaktoren*. Berlin and Köln: Kolhammer.

Grabow, B. (1998) 'Stadtmarketing: Eine Kritische Zwischenbilanz', *Difu Berichte*, 98(1): 2–5.

Granovetter, M. (1985) 'Economic Action and Social Structure: The Problem of Embeddedness', *Amercian Journal of Sociology*, 91: 481–510.

Gray, C. (2006) 'Managing the Unmanageable: The Politics of Cultural Planning', *Public Policy and Administration*, 21(2): 101–113.

Graziano, V. and Trogal, K. (2017) 'The Politics of Collective Repair: Examining Object-Relations in a Postwork Society', *Cultural Studies*, 31(5): 634–658.

Green, N. (2001) *From Factories to Fine Art: A History and Analysis of the Visual Arts Networks in London's East End, 1968–1998*. London: Bartlett School of Planning, University College London.

Greenhalgh, P. (1988) *Ephemeral Vistas. The Expositions Universelles, Great Exhibitions and World's Fairs, 1851–1939*. Manchester: Manchester University Press.

Gripsiou, A. and Bergouignan, C. (2021) 'The Internal Socio-Economic Polarization of Urban Neighborhoods, the Case of Marseille', *Investigaciones Geográficas*, 77: 103–128.

Gu, X. and O'Connor, J. (2014) 'Making Creative Spaces: China and Australia: An Introduction City', *Culture and Society*, 5(3): 111–114.

Guppy, M. (ed) (1997) *Better Places, Richer Communities: Cultural Planning and Local Development – A Practical Guide*. Sydney: Australia Council for the Arts.

Gwee, J. (2009) 'Innovation and the Creative Industries Cluster: A Case Study of Singapore's Creative Industries', *Innovation*, 11(2): 240–252.

Hall, C.M. (1989) 'The Definition and Analysis of Hallmark Tourist Events', *GeoJournal*, 19(3): 263–268.

Hall, P. (1998) *Cities and Civilization: Culture, Innovation, and Urban Order*. London: Weidenfeld & Nicholson.

Hall, P. and Hall, P. (2006) 'Reurbanizing the Suburbs?', *City*, 10(3): 377–392.

Hall, S. (1996) 'Gramsci's Relevance for the Study of Race and Ethnicity', in D. Morley and K.-H. Chen (eds) *Stuart Hall: Critical Dialogues in Cultural Studies*. London: Routledge, 411–440.

Hall, T. and Smith, C. (2005) 'Public Art in the City: Meanings, Values, Attitudes and Roles', *Interventions: Advances in Art and Urban Futures*, 4: 175–179.

Handa, R. (1998) 'Body World and Time: Meaningfulness in Portability', in R. Kronenburg (ed) *Transportable Environments: Theory, Context, Design and Technology*. London: E & FN Spon, 8–17.

Hannigan, J. (1998) *Fantasy City: Pleasure and Profit in the Postmodern Metropolis*. London: Routledge.

Hansen, S. and Flynn, D. (2014) *"Bring Back Our Banksy!" Street Art and the Transformation of Public Space, Lisbon Street Art & Urban Creativity International Conference*. Lisbon: Lisbon University, 3–5 July.

Harland, J. and Kinder, K. (1999) *Crossing the Line: Extending Young People's Access to Cultural Venues*. London: Calouste Gulbenkian Foundation.

Harris, A. (2009) 'Shifting the Boundaries of Cultural Spaces: Young People and Everyday Multiculturalism', *Journal for the Study of Race, Nations and Culture*, 15(2): 187–205.

Harris, J. (2015) 'Embodiment in the Museum – What Is a Museum?', *ICOFOM Study Series*, 43b: 101–115.

Harrison, R. and DeSilvey, C. (2022) *Heritage Futures: Comparative Approaches to Natural and Cultural Heritage Practices*. London: UCL Press.

Harvey, D. (1989) 'From Managerialism to Entrepreneurialism: The Transformation in Urban Governance in Late Capitalism', *Geografiska Annaler. Series B, Human Geography*, 71(1): 3–17.

Hawkes, J. (2001) *The Fourth Pillar of Sustainability: Culture's Essential Role in Public Planning*. Melbourne: Common Ground.

Heebels, B. and van Aalst, I. (2020) 'Creative Clusters in Berlin: Entrepreneurship and the Quality of Place in Prenzlauer Berg and Kreuzberg', *Geografiska Annaler: Series B, Human Geography*, 92(4): 347–363.

Hertzberger, H. (1991) *Lessons for Students in Architecture*. Rotterdam, The Netherlands: Uitgiverij.

Hesmondhalgh, D. (2012) *The Cultural Industries*, 3rd ed. London: SAGE.

Hetherington, K. (1996) 'The Utopics of Social Ordering – Stonehenge as a Museum Without Walls', in S. Macdonald and G. Fyfe (eds) *Theorizing Museums: Representing Identity and Diversity in a Changing World*. Oxford: Berg.

Hewison, R. (1987) *The Heritage Industry. Britain in a Climate of Decline*. London: Routledge.

Higgs, P., Cunningham, S. and Bakhshi, B. (2008) *Beyond the Creative Industries: Mapping the Creative Economy in the United Kingdom*. London: NESTA.

Hillery, G. (1955) 'Definitions of Community: Areas of Agreement', *Rural Sociology*, 20: 111–123.

Hillier, B. and Hanson, J. (1988) *The Social Logic of Space*. Cambridge: Cambridge University Press.

Historic England (2015) *Heritage Counts: The Value and Impact of Heritage*. London: Historic England.

Historic England (2016) *Place branding and Heritage. Heritage Counts*. Evans, G.L., TBR and Pomegranite Seeds. London: Historic England.

Holliss, F. (2015) *Beyond Live/Work. The Architecture of Home-Based Work*. London: Routledge.

Hommels, A. (2005) *Unbuilding Cities – Obduracy in Urban Sociotechnical Change*. Cambridge, MA: MIT Press.

Hope, S. (2017) 'From Community Arts to the Socially Engaged Art Commission', in A. Geffers and G. Moriarty (eds) *Culture, Democracy and the Right to Make Art*. London: Bloomsbury, 203–221.

Hosagrahar, J. (2017) *Culture: At the Heart of the SDGs*. Paris: UNESCO.

Hubbard, P. and Hall, T. (1998) 'The Entrepreneurial City and the New Urban Politics', in T. Hall and P. Hubbard (eds) *The Entrepreneurial City: Geographies of Politics, Regime and Representation*. Chichester: John Wiley & Sons.

Hughes, R. (1991) *The Shock of the New: Art and the Century of Change*. London: Thames and Hudson.

Hutchison, R. and Forrester, S. (1987) *Arts Centres in the UK*. London: PSI.

Hutton, T.A. (2006) 'Spatiality, Built Form, and Creative Industry Development in the Inner City', *Environment and Planning A*, 38(10): 1819–1841.

HWFI (2018) *Hackney Wick & Fish Island Creative Enterprise Zone Workshop*. London: Tom Fleming/We Made That/Regeneris, 18 July.

Impacts08 (2009) *Local Area Studies: 2008 Results*. Liverpool: University of Liverpool and Liverpool John Moores University.

Irvine, M. (2012) 'The Work on the Street, Street Art and Visual Culture', in B. Sandywell and M. Heywood (eds) *The Handbook of Visual Culture*. New York: Berg, 235–278.

Islam, S. and Iversen, K. (2018) 'From "Structural Change" to "Transformative Change": Rationale and Implications', *DESA Working Paper, 155*. New York: United Nations Department of Economic and Social Affairs (DESA).

IT (1969) *International Times Newsletter*, 66, 10–23 October. London.

Jacob, D. and van Heur, B. (2014) 'Taking Matters into Third Hands: Intermediaries and the Organization of the Creative Economy', *Regional Studies*, 49(3): 357–361.

Jacobs, D. (2014) 'Fashion District Arnhem: Creative Entrepreneurs Upgrading a Deprived Neighbourhood', in L. Marques and G. Richards (eds) *Creative Districts Around the World*. Breda: CELTH/NHTV, 74–80.

Jacobs, J. (1961) *The Death and Life of Great American Cities*. Harmondsworth: Penguin.

Jameson, F. (1998) *The Cultural Turn: Selected Writings on the Postmodern, 1983–1998*. Brooklyn: Verso.

Jansson, J. and Power, D. (2010) 'Fashioning a Global City: Global City Brand Channels in the Fashion and Design Industries', *Regional Studies*, 44(7): 889–904.

Jarzombek, M.M. (2004) *Designing MIT: Bosworth's New Tech*. Cambridge, MA: Northeastern University Press.

Jay, M. (2002) 'That Visual Turn', *Journal of Visual Culture*, 1(1): 87–92.

Jeffers, A. and Moriarty, G. (2017) *Culture, Democracy and the Right to Make Art*. London: Bloomsbury.

Jiwa, S., Coca-Stefanik, A., Blackwell, M. and Rahman, T. (2009) 'Light Night: An "Enlightening" Place Marketing Experience', *Journal of Place Management and Development*, 2(2): 154–166.

Jones, J. (2021) 'The More Satirical Street Murals Are, the Less They Resemble Great Art', *The Guardian*, 5 February. www.theguardian.com/artanddesign/2021/feb/05/the-more-satirical-street-murals-are-the-less-they-resemble-great-art (accessed 23 March 2023).

Jones, J. (2023) 'David Hockey: Bigger and Closer Review – An Overwhelming Blast of Passionless Kitsch', *The Guardian*, 21 February.

Jonze, T. (2023) 'A Life in Art. The World According to Hockney', *The Guardian*, 20 February.

Joseph, M. (1998) 'The Performance of Production and Consumption', *Social Text*, 54: 25–61.

Kalantzis, M. and Cope, B. (2016) 'Learner Differences in Theory and Practice', *Open Review of Educational Research*, 3(1): 85–132.

Kanya, A., et al. (2019) 'The Criterion Validity of Willingness to Pay Methods: A Systematic Review and Meta-Analysis of the Evidence', *Social Science & Medicine*, 232: 238–261.

Kaszynska, P. (2017) *The Cultural Value Scoping Project*. London: AHRC, PHF and KCL. www.ukri.org/wp-content/uploads/2021/11/AHRC-291121-AHRCCulturalValueScopingProjectReport.pdf

Kavaratzis, M. (2005) 'Place Branding: A Review of Trends and Conceptual Models', *The Marketing Review*, 5: 329–342.

Kavaratzis, M., Warnaby, G. and Ashworth, G. (eds) (2014) *Rethinking Place Branding: Comprehensive Brand Development for Cities and Regions*. Vienna: Springer.

Kearns, G. and Philo, C. (1993) *The City as Cultural Capital; Past and Present*. Oxford: Pergamon.

Keegan, R. and Kleiman, N. (2005) *Creative New York*. New York: Center for an Urban Future.

Kelly, O. (1984) *Community, Art and the State: Storming the Citadels*. London: Comedia.

Kester, G. (2004) *Conversation Pieces: Community and Communication in Modern Art.* Berkeley: University of California Press.

Khomami, N. (2022) 'David Hockney Joins Immersive Art Trend with New London Show', *The Guardian*, 16 November. www.theguardian.com/artanddesign/2022/nov/16/david-hockney-joins-immersive-art-trend-with-new-london-exhibition

Khovanova-Rubicondo, K.M. (2012) 'Cultural Routes as a Source for New Kind of Tourism Development: Evidence from the Council of Europe's Programme', *International Journal of Heritage in the Digital Era*, 1(1): 83–88.

Kohn, M. (2010) 'Toronto's Distillery District: Consumption and Nostalgia in a Post-Industrial Landscape', *Globalizations*, 7(3): 359–369.

Kornblum, J. (2022) *Marseille, Port to Port.* New York: Columbia University Press.

Kotler, P., Asplund, C., Rein, I. and Heider, D. (1999) *Marketing Places Europe: Attracting Investments, Industries, Residents and Visitors to European Cities, Communities, Regions and Nations.* London: Pearson Education.

Kriedler, J. (2005) *Presentation of Silicon Valley Cultural Initiative. Creative Spaces Study Tour.* Toronto: University of Toronto.

Križnik, B. (2004) 'Transformation of Deprived Urban Areas and Social Sustainability: A Comparative Study of Urban Regeneration and Urban Redevelopment in Barcelona and Seoul', *Urban Izziv*, 29(1): 83–95.

Kronenburg, R. (2003) 'Tadao Ando Karaza Theatre, Japan, 1987–1988', in R. Kronenburg (ed) *Portable Architecture.* London: Routledge.

Kuhlmann, M. (2020) 'What Hamburg's HafenCity Can Learn from the Olympic Games', in G.L. Evans (ed) *Mega-Events, Placemaking, Regeneration and City-Regional Development.* London: Routledge, 140–155.

Kunzmann, K. (2004a) 'Culture, Creativity and Spatial Planning', *Town Planning Review*, 75(4): 383–404.

Kunzmann, K. (2004b) 'Keynote Speech to Intereg III Mid-term Conference. Lille', *Regeneration and Renewal*, 19 November, 2.

Lachmann, R. (1988) 'Graffiti as Career and Ideology', *American Journal of Sociology*, 94(2): 229–250.

Lachmann, R. (1995) 'Review: Crimes of Style: Urban Graffiti and the Politics of Criminality', *Journal of Criminal Justice and Popular Culture*, 3(4): 98–101.

Lacroix, J.-G. and Tremblay, G. (1997) 'The Information Society and Cultural Industries Theory', *Current* Sociology, 45(4): 1–154.

Lacy, S. (1995) *Mapping the Terrain: New Genre Public Art.* Seattle: Bay Press.

Lamont, M. and Aksartova, S. (2002) 'Ordinary Cosmopolitanisms: Strategies for Bridging Racial Boundaries among Working-class Men', *Theory, Culture & Society*, 19(4): 1–25.

Lancaster, H. (2010) *Redefining Places for Art: Exploring the Dynamics of Performance and Location.* Canberra: Australian Research Council.

Landry, C. and Bianchini, F. (1995) *The Creative City.* London: Comedia.

Landry, C. (2000) *The Creative City: A Toolkit for Urban Innovators.* London: Earthscan.

Lane, J. (1978) *Arts Centres-Every Town Should Have One.* London: Paul Elek.

Lane, R. (1998) 'The Place of Industry', *Harvard Architecture Review*, 10: 151–161.

LDA (2003) *Creative London: Vision and Plan.* London: London Development Agency.

Leadbetter, C. (1999) *Living on Thin Air: The New Economy.* London: Penguin.

Leader, D. (2002) *Stealing the Mona Lisa: What Arts Stops us from Seeing.* London: Faber & Faber.

Lee, M. (1997) 'Relocating Location: Cultural Geography, the Specificity of Place and the City of Habitus', in J. McGuigan (ed) *Cultural Methodologies.* London: Sage, 126–141.

Leeson, L. (2018) *Art: Process: Change: Inside a Socially Situated Practice*. London: Routledge.

Leeson, L. (2020) 'Active Energy: Communities Countering Climate Change', *Women Eco Artists Dialog-Magazine & Directory*, 11.

Lefebvre, H. (1968) *Le droit a la ville*. Paris: Anthropos.

Lefebvre, H. (1970/2003) *The Urban Revolution* (trans. R. Bononno). Minneapolis: University of Minnesota Press.

Lefebvre, H. (1974/1991) *The Production of Space* (trans. D. Nicholson). Oxford: Blackwell.

Lefebvre, H. (1984) *Everyday Life in the Modern World* (trans. S. Rabinovitch). New Brunswick: Transaction.

Lefebvre, H. (1991) *Critique of Everyday Life*, Vol. 1 (trans. J. Moore). London: Verso.

Lefebvre, H. (1992) *Rhythmanalysis: Space, Time and Everyday Life* (trans. S. Elden and G. Moore). London: Continuum.

Lefebvre, H. (1996) 'The Right to the City', in E. Kofman and E. Lebas (eds) *Writings on Cities*. Cambridge, MA: Wiley-Blackwell, 147–159.

Lefebvre, H. (2014) *Toward an Architecture of Enjoyment* (ed. L. Stanek; trans. R. Bononno). Minneapolis: University of Minnesota Press.

Lew, A. (2017) 'Tourism Planning and Place Making: Place-Making or Placemaking?', *Tourism Geographies*, 19(3): 448–466.

Lewisohn, C. (2013) *The Canals Project Fanzine*. London.

Ley, D. and Olds, K. (1988) 'Landscape as Festival: World's Fairs and the Culture of Heroic Consumption', *Environment and Planning D: Society and Space*, 6: 191–212.

Lipovetsky, G. (2002) *The Empire of Fashion. Dressing Modern Democracy*. Princeton: Princeton University Press, 89–91.

Littlewood, J. (1964) 'A Laboratory of Fun', *New Scientist*, 14 May, 432–433.

LLDC (2014) *Ten Year Plan Draft V4*. London: Greater London Authority, 9 May.

Lombard, K.-J. (2013) 'Art Crimes: The Governance of Hip Hop Graffiti', *Journal of Cultural Research*, 17(3): 255–278.

Longley, A. and Duxbury, N. (2016) 'Cultural Mapping: Making the Intangible Visible', Special Issue of *City, Culture and Society*, 7(1): 1–7.

Lorente, J.P. (1998) *Cathedrals of Urban Modernity*. Aldershot: Ashgate.

MacCannell, D. (1996) *Tourist or Traveller*? London: BBC Education.

MacKeith, J. (1996) *The Art of Flexibility-Art Centres in the 1990s*. London: Arts Council of England.

Madgin, R. (2021) *Why Do Historic Places Matter? Emotional Attachments to Urban Heritage*. Glasgow: University of Glasgow.

Malik, K. (2022) 'The Web Has Expanded the Reach of Art But Nothing Beats Standing in Front of a Picasso', *The Observer*, 18 September, 48.

Malone, M. (2007) *Andrea Fraser, 'What Do I, as an Artist, Provide*? Mildred Lane Kemper Art Museum. St. Louis: Washington University in St. Louis.

Manning, H. (2023) *My Experience of Coventry City of Culture – A Legacy of Facebook Followers*, Mar 13. https://alisonsatuma.medium.com/my-experience-of-coventry-city-of-culture-a-legacy-of-facebook-followers-9f37fab63476 (accessed 17 March 2023).

Maramotti, L. (2000) 'Connecting Creativity', in N. White and I. Griffiths (eds) *The Fashion Business: Theory, Practice, Image*. New York: Oxford, 91–102.

Marcotte, P. and Bourdeau, L. (2002) 'Tourists' Knowledge of the UNESCO Designation of World Heritage Sites: The Case of Visitors to Quebec City', *International Journal of Arts Management*, 8(2): 4–13.

Markusen, A. and Gadwa, A. (2010) *Creative Placemaking*. Washington, DC: National Endowment for the Arts.

Marquand, D. (2004) *Decline of the Public: The Hollowing Out of Citizenship*. Cambridge: Polity.

Marx, K. (1852) *The Eighteenth Brumaire of Louis Bonaparte Die Revolution*. Moscow: Progress Publishers.

Marx, K. (1973) *Grundrisse* (trans. M. Nicolaus). London: Penguin.

Massey, D. (1999) 'Space-Time, "Science" and the Relationship Between Physical Geography and Human Geography', *Transactions of the Institute of British Geographers*, 24(3): 261–276.

Massey, M. (2014) *Institute of Contemporary Arts, 1946–58*. London: ICA.

Mattern, S. (2014) 'Library as Infrastructure', *Places Journal*, June. https://doi.org/10.22269/140609 (accessed 14 March 2023).

Mathews, S. (2005) 'The Fun Palace: Cedric Price's Experiment in Architecture and Technology', *Technoetic Arts*, 3(2): 73–91.

Matthews, V.L. (2010) *Place Differentiation: Redeveloping the Distillery District, Toronto*. Unpublished PhD Thesis. University of Toronto, Department of Geography.

McAuliffe, C. (2012) 'Graffiti or Street Art? Negotiating the Moral Geographies of the Creative City', *Journal of Urban Affairs*, 34(2): 189–206.

McKinsey & Co (2011) *East London: World-Class Centre for Digital Enterprise*. London: McKinsey & Co, May.

Merrill, S. (2014) 'Keeping It Real? Subcultural Graffiti, Street Art, Heritage and Authenticity', *International Journal of Heritage Studies*, 21(4): 369–389.

Metro-Dynamics (2010) *The Impact of Arts & Culture on the wider Creative Economy*. London: Arts Council England.

Miège, B. (1989) *The Capitalization of Cultural Production*. New York: The Capitalization of Cultural Production.

Miles, S. (2010) *Spaces for Consumption: Pleasure and Placelessness in the Post-Industrial City*. London: Sage.

Miles, M. (2015) *The Symbolic Economy, Limits to Culture: Urban Regeneration vs. Dissident Art*. London: Pluto Press.

Miller-Idriss, C. and Hanauer, E. (2011) 'Transnational Higher Education: Offshore Campuses in the Middle East', *Comparative Education*, 47(2): 181–207.

Montgomery, J. (1995) 'The Story of Temple Bar: Creating Dublin's Cultural Quarter', *Planning, Practice and Research*, 10(2): 135–171.

Montgomery, J. (2013) 'Cultural Quarters and Urban Regeneration', in G. Young and D. Stevenson (eds) *Planning and Culture*. Farnham: Ashgate.

Moore, R. (2022) 'The Day the Music Died', *The Observer*, 7 August, 28–29.

Molina-Morales, F. et al. (2015) 'Formation and Dissolution of Inter-Firm Linkages in Lengthy and Stable Networks in Clusters', *Journal of Business Research*, 68: 1557–1562.

Moriarty, G. (2004) 'The Wedding Community Play Project: A Cross-Community Production in Northern Ireland', in R. Boon and J. Plastow (eds) *Theatre and Empowerment: Community Drama on the World Stage*. Cambridge: Cambridge University Press, 13–32.

Moriarty, G. (2014) *Where Have We Come From? Community Arts to Contemporary Practice*. http://communityartsunwrapped.com/page/2/ (accessed 23 April 2023).

Morris, S. (2022) 'It Will Stay in Our Hearts. Port Talbot Bids Farewell to Banksy Ash Child Mural', *The Guardian*, 29 January. www.theguardian.com/artanddesign/2022/jan/28/port-talbot-prepare-bid-farewell-banksy-mural-seasons-greetings-taken-from-industrial-town-wales (accessed 12 February 2022).

Morrison, W. and West, E. (1986) 'Child Exposure to the Performing Arts: The Implications for Adult Demand', *Journal of Cultural Economics*, 10: 17–24.

Muir, G. and Massey, A. (2014) *ICA London 1946–68*. London: ICA Publishing.

Murtagh, B., Boland, P. and Shirlow, P. (2017) 'Contested Heritages and Cultural Tourism', *International Journal of Heritage Studies*, 23(6): 506–520.

Myerscough, J. (1988) *The Economic Importance of the Arts in Britain Plus Three Reports on The Economic Importance of the Arts in Glasgow, Ipswich and Merseyside*. London: Policy Studies Institute.

NEA (2019) *U.S. Patterns of Arts Participation: A Full Report from the 2017 Survey of Public Participation in the Arts*. Washington. DC: National Endowment for the Arts.

Nermod, O., Lee, N. and O'Brien, D. (2021) *The European Capital of Culture: A Review of the Academic Evidence*. London: Creative Industries Policy and Evidence Centre, London School of Economics and University of Edinburgh.

NESTA (2010) *Beyond Live: Digital Innovation in the Performing Arts*. London: NESTA.

NESTA (2012) *How Big Are the UK's Creative Industries?* London: NESTA.

NFA (2008) *Artists' Studio Provision in the Host Boroughs: A Review of the Potential Impacts of London's Olympic Project*. National Federation of Artists' Studio Providers. London: NFA, December.

Nichols Clark, T. (ed) (2011) *City as Entertainment Machine*. Lanham, MD: Lexington Books.

Nitzsche, S. (2020) Review: 'Routledge Handbook of Graffiti and Street Art', J. Ross (ed.), *Global Hip Hop Studies*, 1(1): 162–165.

Noonan, D. (2013) 'How US Cultural Districts Shape Neighbourhoods', *Cultural Trends*, 22(3–4): 203–212.

Norberg-Schulz, C. (1979) *Genius Loci: Towards a Phenomenology of Architecture*. New York: Rizzoli.

nVision (2006) *A Life of Leisure: Executive Summary*. London: The Future Foundation.

OBU [Oxford Brookes University] (2003) *Townscape Heritage Initiative Schemes Evaluation: Interim Report Summary*. London: Heritage Lottery Fund.

O'Callaghan, B. (2023) 'The Gallery Without Walls. Season Two: The State We're In', *The Guardian*, 3 February, 1–14.

O'Connor, J. (2007) *The Cultural and Creative Industries: A Review of the Literature*. London: Arts Council England.

O'Connor, J. (2010) *The Cultural and Creative Industries: A Review of the Literature*. Newcastle: Creativity, Culture and Education.

Oevermann, H., et al. (2022) 'Heritage Requires Citizens' Knowledge: The COST Place-Making Action and Responsible Research', in H. Mieg (ed) *The Responsibility of Science*, Vol. 27. Vienna: Springer, 233–255.

O'Neil, D. (2006) *What Is Culture?* www.palomar.edu/anthro/culture/culture_1. htm#:~:text=While%20human%20societies%20and%20cultures,people%20interacting%20with%20each%20other (accessed 13 May 2023).

Ottati, G.D. (2014) 'A Transnational Fast Fashion Industrial District: An Analysis of the Chinese Businesses in Prato', *Cambridge Journal of Economics*, 38(5): 1247–1274.

Page, J. (2020) *Industries Without Smokestacks. Firm Characteristics and Constraints to Growth*. AGR Working Paper #23. Washington, DC: Brookings Institute.

Pain, R., Whitman, G., Milledge, D. and Lune Rivers Trust (2012) *Participatory Action Research Toolkit: An Introduction to Using PAR as an Approach to Learning, Research and Action*. www.dur.ac.uk/resources/beacon/PARtoolkit.pdf

Palmer, R. (2004) *European Cities and Capitals of Culture: Study Prepared for the European Commission*. Brussels: Palmer/RAE Associates.

Palmer, R. and Richards, G. (2010) *Eventful Cities: Cultural Management and Urban Revitalisation*. London: Butterworth.

Pandolfi, V. (2015) *Fashion and the City: The Role of the 'Cultural Economy' in the Development Strategies of Three Western European Cities*. Eburon: Tilburg University.

Park, H.Y. (2013) *Heritage Tourism*. London: Routledge.

Partridge, J. (2022) 'Search 'office'. Rooftop Walks and a Pool set to Lure Google Staff into its New HQ', *The Guardian*, 2 July.

Paterson, M. (2006) *Consumption and Everyday Life*. London: Routledge.

Peck, J. (2005) 'Struggling with the Creative Class', *International Journal of Urban and Regional Research*, 29(4): 740–770.

Pick, J. (1997) *The Arts Industry*. Gresham College Lecture. London: City University, 14 April.

Pick, J. and Anderton, M. (2013) *Building Jerusalem. Art, Industry and the British Millennium*. London: Routledge.

Pickering, J. (1999) 'Designs on the City', in J. Downey and J. McGuigan (eds) *Technocities*. London: Sage.

Pine, B.J. and Gilmore, J.H. (1998) 'Welcome to the Experience Economy', *Harvard Business Review*, 97–105.

Plaza, B. (2009) 'Bilbao's Art Scene and the 'Guggenheim effect' Revisited', *European Planning Studies*, 17(11): 1711–1729.

Plaza, B. (2015) 'Culture-Led City Brands as Economic Engines: Theory and Empirics', *Annals of Regional Science*, 54(2015): 179–196.

Plaza, B., González-Casimiro, P., Moral-Zuazo, P. and Waldron, C. (2015) 'Culture-Led City Brands as Economic Engines: Theory and Empirics', *The Annals of Regional Science*, 54(1): 179–196.

Plaza, B., Tironi, M. and Haarich, S.N. (2009) 'Bilbao's Art Scene and the "Guggenheim Effect" Revisited', *European Planning Studies*, 17(11): 1711–1729.

Plieninger, T., Dijks, S., Oteros-Rozas, E. and Bieling, C. (2013) 'Assessing, Mapping, and Quantifying Cultural Ecosystem Services at Community Level', *Land Use Policy*, 33: 118–129.

Ponzi, D. and Nastasi, M. (2016) *Starchitecture: Actors and Spectacles in the Global City: Scenes, Actors, and Spectacles in Contemporary Cities*. New York: Monacelli Press.

Ponzini, D. and Nastasi, M. (2011) *Starchitecture*. Venice: Allemandi.

Powell, H. and Marrero-Guillamon, I. (eds) (2012) *The Art of Dissent*. London: Marshgate Press.

Pratt, A.C. (2004) 'Creative Clusters: Towards the Governance of the Creative Industries Production System?', *Media International Australia*, 112(1): 50–66.

Prentice, R. (1993) *Tourism and Heritage Attractions*. London: Routledge.

Presence, S. (2019) 'Britain's First Media Centre: A History of Bristol's Watershed Cinema, 1964–1998', *Historical Journal of Film, Radio and Television*, 39(4): 803–831.

Prigge, W. (2008) 'Reading the Urban Revolution: Space and Representation', in K. Goonewardena, S. Kipfer, R. Milgrom and C. Schmid (eds) *Space, Difference, Everyday Life Reading Henri Lefebvre*. New York: Routledge, 46–61.

Protherough, R. and Pick, J. (2002) *Managing Britannia. Culture and Management in Modern Britain*. Denton: Brynmill Press.

Purcell, M. (2013) 'Possible Worlds: Henri Lefebvre and the Right to the City', *Journal of Urban Affairs*, 36(1): 141–154.

Rantisi, N. (2002) 'The Local Innovation System as a Source of Variety: Openness and Adaptability in New York City's Garment District', *Regional Studies*, 36(6): 587–602.

Raustiala, K. and Sprigman, C. (2006) 'The Piracy Paradox: Innovation and Intellectual Property in Fashion Design', *Virginia Law Review*, 92(8): 1687–1777.

Read, H. (1964) *Contemporary British Art*. London: Pelican.

Read, S. (2017) *Cinderella River: The Evolving Narrative of the River Lee*. London: Middlesex University.

Redström, J. (2006) 'Towards User Design? On the Shift from Object to User as the Subject of Design', *Design Studies*, 27(2): 123–139.

Richards, G. (ed) (1996) *Cultural Tourism in Europe*. Wallingford: CAB International.

Richards, G. (2017) 'From Place Branding to Placemaking: The Role of Events', *International Journal of Event and Festival Management*, 8(1): 8–23.

Richards, G. (2019) 'Creative Tourism: Opportunities for Smaller Places?', *Tourism and Management Studies*, 15: 7–10.

Richards, G., de Brito, M. and Wilks, L. (eds) (2013) *Exploring the Social Impacts of Events*. London: Routledge.

Richards, G. and Marques, L. (2015) 'Exploring Creative Tourism: Introduction', *Transfusion*, 4(2): 1–11.

Richards, G. and Wilson, J. (2006) 'Developing Creativity in Tourist Experience: A Solution to the Serial Reproduction of Culture', *Tourism Management*, 27(6): 1209–1223.

Riding, A. (2007) 'The Industry of Art Goes Global', *The New York Times*, 28 March.

Rissola, G., Bevilacqua, C., Monardo, B. and Trillo, C. (2019) *Place-Based Innovation Ecosystems: Boston-Cambridge Innovation Districts (USA)*. Luxembourg: Publications Office of the European Union.

Roche, M. (2000) *Mega-events and Modernity. Olympics and Expos in the Growth of Global Culture*. London: Routledge.

Roche, M. (2003) 'Mega-events, Time and Modernity: On Time Structures in Global Society', *Time & Society*, 12(1): 99–126.

Roodhouse, S. (2010) *Cultural Quarters. Principles and Practice*, 2nd ed. Bristol: Intellect.

Ross, J. (2016) *Routledge Handbook of Graffiti and Street Art*. New York: Routledge.

Ross, J. and Wright, B. (2014) '"I've Got Better Things to Worry About": Police Perceptions of Graffiti and Street Art in a Large Mid-Atlantic City', *Police Quarterly*, 17(2): 176–200.

Roth, M. (2015) *Victorian Futures*. London: Chelsea College of Art, 14–15 May.

RSA/UCL (2018) *Cities of Making: 03 London*. London: RSA.

Ryan, R.L. (2011) 'The Social Landscape of Planning: Integrating Social and Perceptual Research with Spatial Planning Information', *Landscape and Urban Planning*, 100: 361–363.

Said, E. (1979) *Orientalism*. New York: Knopf Doubleday.

Said, E. (1994) *Culture and Imperialism*. London: Vintage.

Santagata, W. (2004) 'Creativity, Fashion and Market Behavior', in D. Power and A. Scott (eds) *Cultural Industries and the Production of Culture*. London: Routledge, 75–90.

Sassoon, D. (2006) *The Culture of the Europeans: From 1800 to the Present*. London: Harper Press.

Saxenian, A. (1994) *Regional Advantage: Culture and Competition in Silicon Valley and Route 128*. Cambridge, MA: Harvard University Press.

Schouvaloff, A. (ed) (1970) *Place for the Arts*. North West Arts Association. Liverpool: Seel House Press.

Schubert, K. (2000) *The Curator's Egg. The Evolution of the Museum Concept from the French Revolution to the Present Day*. London: One-Off Press.

Scott, A.J. (1997) 'The Cultural Economy of Cities', *International Journal of Urban and Regional Research*, 21(2): 323–339.

Scott, A.J. (2000) *The Cultural Economy of Cities*. London: Sage.

Scott, A.J. (ed) (2001) *Global City-Regions: Trends, Theory, Policy*. New York: Oxford University Press.

Scott, A.J. (2014) 'Beyond the Creative City: Cognitive-Cultural Capitalism and the New Urbanism', *Regional Studies*, 48(4): 565–578.

Sennett, R. (1997) *Carne y piedra. El cuerpo en la civilización occidental*. Madrid: Alianza.

Sennett, R. (1998) *Raoul Wallenberg Lecture*. Michigan: University of Michigan.

Sennett, R. (2000) 'Reflections on the Public Realm', in G. Bridge and S. Watson (eds) *A Companion to the City*. Blackwell: Oxford, 380–387.

Sennett, R. (2008) *The Public Realm*. https://intensificantvidesnervioses.wordpress.com/2013/08/28/the-public-realm-richard-sennett (accessed 20 March 2023).

Servais, E., Quartier, K. and Vanrie, J. (2022) 'Experiential Retail Environments in the Fashion Sector', *Fashion Practice, The Journal of Design, Creative Process & the Fashion Industry*, 14(3): 449–468.

Shaffi, S. (2023) 'It's the Opposite of Art', *The Guardian*, 23 January, 8–9.

Shaw, P., et al. (2006) *Arts Centres Research*. London: Arts Council of England.

Shaw, S.J. (2012) 'Faces, Spaces and Places, Social and Cultural Impacts of Street Festivals in Cosmopolitan Cities', in S. Page and J. Connell (eds) *Routledge Handbook of Event Studies*. London: Routledge, 401–414.

Shaw, S.J. and Macleod, N. (2000) 'Creativity and Conflict: Cultural Tourism in London's City Fringe', *Tourism Culture and Communication*, 2: 165–175.

Shedroff, N. (2001) *Experience Design 1*. Thousand Oaks: New Riders Publications.

Sherwood, H. (2023) 'From Rebel to Fame: Public Gets Access to David Bowie Archive', *The Guardian*, 23 February, 11.

Shorthouse, J. (2004) 'Nottingham's de Facto Cultural Quarter: The Lace Market, Independents and a Convivial Ecology', in D. Bell and M. Jayne (eds) *City of Quarters*. Aldershot: Ashgate, 149–162.

Shove, E., Pantzar, M. and Watson, M. (2012) *The Dynamics of Social Practice: Everyday Life and How It Changes*. London: SAGE.

Shove, E., Watson, M., Hand, M. and Ingram, J. (2007) *The Design of Everyday Life*. New York: Berg.

Simetrica, J. (2021a) *Heritage and the Value of Place*. London: Historic England.

Simetrica, J. (2021b) *Culture and Heritage Capital Evidence Bank – Economic Values Database*. London: DCMS.

Simmel, G. (1950) 'The Metropolis and Mental Life', in K.H. Wolff (ed) *The Sociology of Georg Simmel*. Glencoe, IL: The Free Press, 409–424.

Simmel, G. (1958) 'The Ruin', *The Hudson Review*, 11(3): 379–385.

Simmel, G. (1971) *On Individuality and Social Forms: Selected Writings*. Chicago: University of Chicago Press.

Simmie, J. (2006) 'Do Clusters or Innovation Systems Drive Competitiveness?', in B. Asheim, P. Cooke and R. Martin (eds) *Clusters and Regional Development*. London: Routledge, 164–187.

Simmie, J., Carpenter, J., Chadwick, A. and Martin, R. (2008) *History Matters: Path Dependence and Innovation in British Cities*. London: NESTA.

Simpson, M. (2018) 'Heritage: Nonwestern Understandings', in S. Lopez Varela (ed) *The Encyclopaedia of Archaeological Sciences*. London & New York: John Wiley & Sons, 1–5.

Smith, A. (2016) *Events in the City: Using Public Spaces as Event Venues*. Abingdon: Routledge.

Smith, J. (2006) *Uses of heritage*. London & New York: Routledge.

Smith, V.L. (ed) (1977) *Hosts and Guests: The Anthropology of Tourism*. Oxford: Basil Blackwell.

Smith, A., Vodicka, G., Colombo, A., Lindstrom, K.N., McGillivray, D. and Quinn, B. (2021) 'Staging City Events in Public Spaces: An Urban Design Perspective', *International Journal of Event and Festival Management*, 12(2): 224–239.

Soja, E.W. (1996) *Thirdspace: Journeys to Los Angeles and Other Real-and-Imagined Places*. London: Wiley-Blackwell.

Solnit, R. (2001) *Hollow City: The Siege of San Francisco and the Crisis of American Urbanism*. Brooklyn: Verso.

Southern, R. (1962) *The Seven Ages of the Theatre*. London: Faber & Faber.

Stanek, L. (2014) 'Introduction', in H. Lefebvre (ed) *Toward an Architecture of Enjoyment* (trans. R. Bononno). Minneapolis: University of Minnesota Press.

Stanek, L. and Schmid, C. (eds) (2014) *Urban Revolution Now: Henri Lefebvre in Social Research and Architecture*. London: Routledge.

Stark, P. (1984) *The Unplanned Arts Center as a Base for Planned Growth in Arts Provision*. London: City University.

Stern, M.J. and Seifert, S. (2007) *Cultivating "Natural" Cultural Districts*. Philadelphia: Penn University The Social Impact of the Arts Programme.

Stern, M.J. and Seifert, S. (2010) 'Cultural Clusters: The Implications of Cultural Assets Agglomeration for Neighbourhood Revitalization', *Journal of Planning Education and Research*, 29: 262–279.

Stoker, G. and Mossberger, K. (1994) 'Urban Regime Theory in Comparative Perspective', *Environment and Planning C: Government and Policy*, 12: 195–212.

Store Smith, J. (1972) 'Social Aspects', in A. Briggs (ed) *Victorian People: A Reassessment of Persons and Themes, 1851–1867*. Chicago: University of Chicago Press.

Storper, M. and Scott, A.J. (2009) 'Rethinking Human Capital, Creativity and Urban Growth', *Journal of Economic Geography*, 9: 147–167.

Sudjic, D. (1993) *The 100 Mile City*. London: Flamingo.

Sudjic, D. (2005) *The Edifice Complex: How the Rich and Powerful Shape the World*. New York: Penguin.

TAC (1988) *No Vacancy: A Cultural Facilities Policy for the City of Toronto*. Toronto: Toronto Arts Council.

TBR and Cities Institute (2011) *The Art of the Possible: Using Secondary Data to Detect Social and Economic Impacts from Investments in Culture and Sport: A Feasibility Study*. CASE Programme. London: DCMS.

TBR, Evans, G.L. and NEF (2016) *The Contribution of Culture to Placeshaping*. CASE Programme. London: DCMS.

Thomas, D. (2004) 'Mcfashion Design', *Newsweek*, 30 November. www.newsweek.com/mcfashion-design-124199 (accessed 2 March 2023).

Thompson, N. (2012) *Living as Form: Socially Engaged Art from 1991–2011*. New York: Creative Time Books/MIT Press.

Tunbridge, J. and Ashworth, G. (1997) 'Dissonant Heritage. The Management of the Past as a Resource in Conflict', *Annals of Tourism Research*, 24(2): 496–498.

UCLG (United Cities and Local Governments) (2004) *Agenda 21 for Culture. An Undertaking by Cities and Local Governments for Cultural Development*. Barcelona: UCLG/City of Barcelona.

UNESCO (2001) *International Round Table on 'Intangible Cultural Heritage – Working Definitions*. UNESCO: Turin.

UNESCO (2009) *Right of Everyone to Take Part in Cultural Life*, art. 15, para. 1(a), of the International Covenant on Economic, Social and Cultural Rights. Geneva: Committee on Economic, Social and Cultural Rights.

University of Hull (2021) *The Impacts of Hull UK City of Culture 2017. Main Evaluation Findings and Reflections*. Hull: University of Hull.

Unwin, T. (2000) 'A Waste of Space? Towards a Critique of the Social Production of Space', *Transactions of the Institute of British Geographers*, 25(1): 11–29.

Urry, J. (1995) *Consuming Places*. London: Routledge.

van Heur, B. (2010) *Creative Networks and the City. Towards a Cultural Political Economy of Aesthetic Production*. Verlag, Bielfield: Transcipt.

van Heur, B., Evans, G.L., de Wilde, R. and Peters, P. (2011) *VIA2018: Maastricht as Knowledge and Learning Region*. Report to the VIA2018 Project Office in Preparation for the Bid Book Maastricht European Capital of Culture 2018. Maastricht: Maastricht University.

Vecchi, A. and Evans, G.L. (2018) *The Changing Nature of the Fashion Industry and Its Impact on Place-Making*, CLUSTERING 2018, 3rd International Conference on Clusters and Industrial Districts. Valencia: University of Valencia, May 24–25.

Vercelloni, M. (2014) *Cluster Pavilions EXPO Milano 2015*. Milan: Internigo.

Vervaeke, M. and Lefebvre, B. (2002) 'Design Trades and Inter-firm Relationships in the Nord-Pas de Calais Textile Industry', *Regional Studies*, 36(6): 661–673.

Verwijnen, J. and Lehtovuori, P. (eds) (1999) *Managing Urban Change*. Helsinki: University of Art and Design Helsinki.

Wainwright, O. (2013) 'Olympic Legacy Murals Met with Outrage by London Street Artist's', *The Guardian*, 6 August. www.theguardian.com/artanddesign/2013/aug/06/olympic-legacy-street-art-graffiti-fury (accessed 3 August 2014).

Wainwright, O. (2022) 'Why Not Go Full Vegas? Crass Reinvention of Central London', *The Guardian*, 29 October, 23.

Wallace, N. (1993) 'Introductory Paper to the Symposium on the Future of London Arts Centres', *Drill Hall*, 13 September. London: London Arts Board.

Wallerstein, I. (1991) 'World System Versus World-Systems: A Critique', *Critique of Anthropology*, 11(2): 189–194.

Walmsley, B. (2012) 'Towards a Balanced Scorecard: A Critical Analysis of the Culture and Sport Evidence (CASE) Programme', *Cultural Trends*, 21(4): 325–334.

Ward, S. (1998) *Selling Places: The Marketing and Promotion of Towns and Cities 1850–2000*. London: E & FN Spon.

Warren, S. and Jones, P. (2015) *Creative Economies, Creative Communities: Rethinking Place, Policy and Practice*. London: Routledge.

Waterton, E. and Watson, S. (2015) *The Palgrave Handbook of Contemporary Heritage Research*. London: Palgrave.

Weightman, G. (1992) *Bright Lights, Big City: London Entertained 1830–1950*. London: Collins & Brown.

White, M. (1998) 'Healthy Living Centres – The Arts Remedy', *Mailout*, December/January.

Wieditz, T. (2007) 'Liberty Village: The Makeover of Toronto's King and Dufferin Area', *Research Bulletin #2*. Centre for Urban and Community Studies. Toronto: University of Toronto.

Williams, E. and Currid-Halkett, E. (2011) 'The Emergence of Los Angeles as a Fashion Hub: A Comparative Spatial Analysis of the New York and Los Angeles Fashion Industries', *Urban Studies*, 48(14): 3043–3066.

Williams, R. (1961) *The Long Revolution*. London: Pelican.

Wilson, J.Q. and Kelling, G.L. (1982) 'Broken Windows: The Police and Neighborhood Safety', *Atlantic Monthly*, 249: 29–38.

Winship, L. (2023) 'Music and Movement Combine to Make Mesmerising Show'. Dance Review of Tom Dale Company, The Place, London. *The Guardian*, 23 March, 19.

Wolff, J. (1981) *The Social Production of Art*. London: Macmillan.

Wolfson, S. (2023) '"It Was a Perfect Storm. I Was Dressing Tupac"; Tommy Hillfinger on Fashion, Race and Aspiration', *The Guardian*, 20 February. www.theguardian.com/fashion/2023/feb/20/tommy-hilfiger-interview-fashion-brand-history (accessed 16 March 2023).

Worpole, K. (2013) *Contemporary Library Architecture: A Planning and Design Guide*. London: Routledge.

Wu, W. (2005) *Dynamic Cities and Creative Clusters*. Working Paper No. 3509. Washington, DC: World Bank Policy Research.

Zenker, S. (2011) 'How to Catch a City? The Concept and Measurement of Place Brands', *Journal of Place Management and Development*, 4(1): 40–52.

Zenker, S. and Beckmann, S. (2013) 'Measuring Brand Image Effects of Flagship Projects for Place Brands: The Case of Hamburg', *Journal of Brand Management*, 20: 642–655.

Zieleniec, A. (2016) 'The Right to Write the City: Lefebvre and Graffiti', *Urban Environment*, 10: 1–17.

Zukin, S. (1982) *Loft Living: Culture and Capital in Urban Change*. New York: Rutgers University Press.

Zukin, S. (1995) *Culture of Cities*. Cambridge, MA: Blackwell.

Index

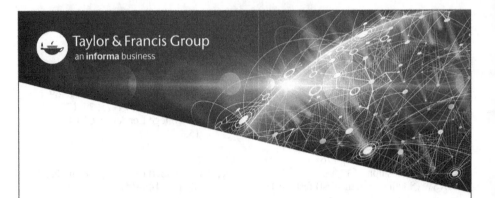

Taylor & Francis Group
an **informa** business

Taylor & Francis eBooks

www.taylorfrancis.com

A single destination for eBooks from Taylor & Francis
with increased functionality and an improved user
experience to meet the needs of our customers.

90,000+ eBooks of award-winning academic content in
Humanities, Social Science, Science, Technology, Engineering,
and Medical written by a global network of editors and authors.

TAYLOR & FRANCIS EBOOKS OFFERS:

A streamlined
experience for
our library
customers

A single point
of discovery
for all of our
eBook content

Improved
search and
discovery of
content at both
book and
chapter level

REQUEST A FREE TRIAL

support@taylorfrancis.com

 Routledge
Taylor & Francis Group

 CRC Press
Taylor & Francis Group

Printed in the United States
by Baker & Taylor Publisher Services

Printed in the United States
by Baker & Taylor Publisher Services